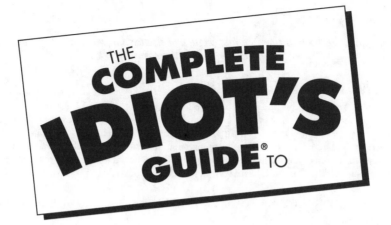

THE COMPLETE **IDIOT'S** GUIDE® TO

Dangerous Diseases and Epidemics

by David Perlin, Ph.D., and Ann Cohen

A
ALPHA
A Pearson Education Company

*We dedicate this book to the men and women of science and medicine who have made
the conquest of disease their lives' work.*

Copyright © 2002 by David Perlin and Ann Cohen

THE COMPLETE IDIOT'S GUIDE TO and Design are registered trademarks of Pearson Education, Inc.

International Standard Book Number: 0-02-864359-3
Library of Congress Catalog Card Number: 2002106338

04   03   02   8   7   6   5   4   3   2   1

Interpretation of the printing code: The rightmost number of the first series of numbers is the year of the book's printing; the rightmost number of the second series of numbers is the number of the book's printing. For example, a printing code of 02-1 shows that the first printing occurred in 2002.

Printed in the United States of America

Note: This publication contains the opinions and ideas of its authors. It is intended to provide helpful and informative material on the subject matter covered. It is sold with the understanding that the authors and publisher are not engaged in rendering professional services in the book. If the reader requires personal assistance or advice, a competent professional should be consulted.

For marketing and publicity, please call: 317-581-3722

The publisher offers discounts on this book when ordered in quantity for bulk purchases and special sales.

For sales within the United States, please contact: Corporate and Government Sales, 1-800-382-3419 or corpsales@pearsontechgroup.com

Outside the United States, please contact: International Sales, 317-581-3793 or international@pearsontechgroup.com

Publisher: *Marie Butler-Knight*
Product Manager: *Phil Kitchel*
Managing Editor: *Jennifer Chisholm*
Acquisitions Editor: *Eric Heagy*
Development Editor: *Jennifer Moore*
Production Editor: *Katherin Bidwell*
Copy Editor: *Cari Luna*
Illustrator: *Jody Schaeffer*
Cover/Book Designer: *Trina Wurst*
Indexer: *Lisa Wilson*
Layout/Proofreading: *Angela Calvert, John Etchison, Rachel Haynes, Vicki Keller*

Contents at a Glance

Contents

Foreword

If you are contemplating reading this book, you're anything but an idiot. You obviously want to increase your knowledge of a most riveting and important subject—infectious diseases. Our knowledge of factors surrounding virulence and immunity has increased an extraordinary degree over the past few decades. That in itself is an extremely interesting story and worth reading about. It's also important to know about infectious diseases to protect yourself and others around you from exposure to infection. And in these days of extensive travel and potential bioterrorism, it's even more important to know what to expect and do based on a deeper understanding of how organisms spread and invade, how we might avoid them, and how our immune system protects us.

For those working with infectious diseases, these are exciting times. We have made extraordinary advances in our understanding of how microorganisms work; how they multiply, invade, produce toxins, and evade our immune responses and our attempts to treat them.

Despite this knowledge, public health officials, physicians, and the general public have every reason to be worried. People, especially children, are dying from infectious diseases at frightening rates because of poor nutrition, crowding, ignorance, and lack of preventive measures and medical care. Poverty is a major reason we still see large numbers of people in certain countries suffer from disease. Refugees are notoriously susceptible to epidemics because of crowding and malnutrition. The numbers of people with AIDS, whose immune systems are slowly destroyed by the HIV, are rapidly increasing in the developing world, especially sub-Saharan Africa and Asia. To make matters worse, governments are developing biological weapons to fight wars—and they are using those weapons. More often than not, disease is the result of man-made problems in our world.

Nevertheless, fields like molecular biology and molecular epidemiology are continuing to clarify how some diseases spread so that we can use preventative measures to stop them. And some of these measures are very easy to follow. This has clearly been demonstrated in hospital-acquired infections. It's ironic that the most effective prevention for stopping the spread of hospital-acquired infection remains the simple method of washing hands between seeing and touching each patient.

Educating ourselves about infectious disease and its risk is critical. A positive development early in the AIDS epidemic was that people at risk for HIV infection became educated about the disease and organized to educate others and influence various agencies to respond appropriately with programs of prevention, treatment, and research. In addition, the people at risk in many areas of the developed world sought and obtained funds to support these efforts. There are other examples of a well-educated public taking a hand in the control of a specific disease such as breast, prostate, and lung cancer.

This book by Perlin and Cohen presents information on infectious diseases in a clear and lively format that can only promote intervention by the public in control of infectious diseases. An educated public can influence those who are in positions to legislate and fund control measures. With all our knowledge about infections and their consequences, it is sad that infections remain the leading cause of death worldwide. We all need to work to correct this.

Donald Armstrong, MD, MACP
Consultant, Infectious Disease Service
Memorial Sloan-Kettering Cancer Center
Professor of Medicine
Cornell University Medical College

Introduction

You're not feeling all that well—you've been tired the past couple of days, and now you're beginning to feel achy and have a slight fever with a touch of nausea. You figure you either finally caught the cold your four-year-old brought home with her from day care last week or got a touch of the flu that your assistant came down with on Monday. Then again, you ate at that new sushi restaurant for lunch yesterday, and you know what they say about the hazards of eating raw fish .... And it's only been two weeks since you returned from your trip of a lifetime—an African safari—where even though you were told not to drink the water, you did.

Now change the details: Replace the sick colleague with a mosquito bite. Instead of eating sushi, make it undercooked hamburger. Make the African safari a trip to the local swimming hole or unprotected sex.

The point of this example is to demonstrate that infectious diseases—including the common cold and flu, food-borne illnesses, and tropical, waterborne, and sexually transmitted diseases—can be "caught" in any number of situations as you go about your daily life. This book is intended to inform you about those diseases, helping you to understand how the body can survive in such a seemingly hostile environment and providing you with practical knowledge about a number of the most common diseases, as well as some that are more rare and exotic.

If you find, after reading this book, that you can entertain your buddies for hours with horror stories about great epidemics of the past, or scare the pants off of them with tales of emerging infections that are resistant to all known drugs, that's a bonus.

How This Book Is Organized

We've organized this book into four parts. In the first part, **"Understanding the Enemy: Infection and Disease,"** we describe the nature of infectious organisms—what they look like, where they live, how they work—and how the body and doctors fight them off. You'll also learn about people's early battles with disease—they won some and lost many, and left others for us to contend with. We encourage you to read the first five chapters before moving on to read about specific diseases, as they lay the groundwork for the rest of the book.

Part 2, "Raging Epidemics" looks at major disease outbreaks that people around the world are dealing with today: AIDS, tuberculosis, malaria, sexually transmitted diseases, and hepatitis. Most of these diseases have already killed millions of people, and have the potential to kill millions more if left unchecked.

The diseases described in **Part 3, "Hot Viruses and Other Killer Bugs,"** range from a mild case of athlete's foot to a terrifying outbreak of Ebola in Zaire. We grouped the diseases into chapters based on how they attack the body, who their victims usually are, or where they thrive. For example, many bugs, such as those that cause river blindness and yellow fever, love the heat and moisture of the tropics and are described in Chapter 16.

In the final part, **"Scourges of Our Own Creation,"** we look at the bugs that we have somehow created or given the opportunity to thrive. Ranging from weaponized anthrax to drug-resistant staph, these organisms can help us to realize that we have a lot to learn about infectious organisms and how our body fights them before we can declare victory on infectious disease.

Along the Way ...

Throughout the chapters you'll find the following boxes:

 Disease Diction _____

Medical jargon and scientific terms, simplified.

Infectious Knowledge

You'll be spreading this trivia faster than an outbreak of chickenpox.

 **Potent Fact** _____

An apple a day keeps the doctor away? Probably not. But following the tips in these boxes just might keep disease at bay.

 **Antigen Alert** _____

Antigens are like the bugs' fingerprints—once the body locates and traces the prints, it can take action against the infection. Like antigens, these boxes will alert you to potential hazards when dealing with infectious diseases.

Acknowledgments

Dr. Perlin would like to thank his family, Amy, Josh, and Dan, for their patience, support, and encouragement over many long weekends and nights, which helped greatly in the writing of this book. Ms. Cohen would like to thank her family and friends, particularly Len, Sandy, David, Carol, and Caroline, who supported and encouraged her as well. Dr. Perlin and Ms. Cohen also thank Susan Barry, without whom there would be no book, and Marcia Layton Turner, whose fast review of each chapter was tremendously helpful during the writing process.

Trademarks

Part 1

Understanding the Enemy: Infection and Disease

When it comes to infectious diseases, the enemy is everywhere—even inside us. In this part of the book you'll find out just what that enemy is, including the forms it takes and its multiple means of attack. You'll find out about humankind's centuries-long struggle with diseases, and the many small (and some not so small!) steps we've taken to identify, treat, prevent, and sometimes even eradicate infectious diseases.

Ingenuity alone isn't responsible for the successes we've had against the enemy, though; our bodies' immune systems have developed amazing techniques for preventing infectious diseases and, if that doesn't work, for fighting off diseases once they've taken hold. Of course, the good guys don't always win, and the final chapter in this section tells the stories of the great epidemics that terrorized, mutilated, and killed millions of people in our not-so-distant past.

Catching On to Infectious Diseases

In This Chapter

- ◆ Understanding infectious diseases
- ◆ Outbreaks, epidemics, and pandemics
- ◆ How infectious diseases are caused and spread
- ◆ Victories and setbacks
- ◆ Major challenges in the war against disease

In 1970, the United States Surgeon General William Stewart said that we were "ready to close the book on infectious diseases as a major health threat." His optimism resulted from a firm belief that modern medicine, antibiotics, vaccination, and sanitation methods would soon defeat most infectious diseases.

Unfortunately, his optimism was premature. According to the World Health Organization (WHO), an international agency based in Geneva, Switzerland, that gathers information on and helps to prevent many types of diseases, one third of all deaths worldwide each year are due to an infectious disease. The leading causes of deaths from infectious diseases are acute respiratory infections, tuberculosis, diarrhea, HIV/AIDS, and malaria.

Before we get into why infectious diseases are still a major problem, let's first take a look at what infectious diseases are.

What Is an Infectious Disease?

A disease is a condition of the whole body or part of the body that impairs normal functioning. Some diseases are caused by genetic mutations and heredity, like cystic fybrosis and hemophilia. Other diseases are promoted by living conditions such as poverty, or lifestyle decisions, such as smoking and bad nutrition, which may result in lung cancer, heart disease, or diabetes.

Infectious diseases are illnesses caused by microorganisms (microbes), viruses, or other biological agents, which are too small to be seen without the aid of a microscope. These disease-causing agents can be spread to others, and they can cause serious illness or death. Infectious diseases can be spread in a frightening variety of ways, including the following:

- Through the air
- Person to person
- By touching infectious material
- From a healthy "carrier" of the disease-causing organism
- By animals, including household pets
- By insects such as mosquitoes, fleas, and ticks
- From infected water or food

Antigen Alert

A healthy person's mouth has 10 to 15 million microorganisms in a milliliter of saliva. A sick person's mouth may contain 150 million microorganisms in a milliliter of saliva.

The manner in which a disease is spread depends on the particular organism causing the disease. For example, the virus that causes AIDS isn't spread through the air or from mosquito bites. It is spread through person-to-person sexual contact.

Infectious diseases cause one third of the world's deaths each year. The most common killers are listed in the bar graph.

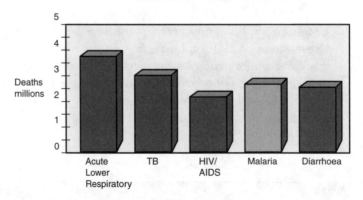

Infectious and Parasitic Diseases

Deaths millions

Acute Lower Respiratory | TB | HIV/AIDS | Malaria | Diarrhoea

Just as not all diseases are infectious, not all infections cause disease. If the organism causing an infection is quickly recognized and killed by the body's immune system, for example, it will not have enough time to cause disease.

Like most things created by nature, infectious diseases are wildly diverse, ranging from the common cold to the exotic Ebola. Infectious diseases also serve as examples of nature's bounty, for there are hundreds of different organisms responsible for hundreds of diseases.

Several different kinds of biological entities are responsible for causing infectious diseases; they are bacteria, fungi, viruses, parasites, protozoans, and prions. Some of these organisms, like bacteria and fungi, can live on their own. Others, like viruses, cannot survive or multiply without a *host*.

Disease Diction

An **infectious disease** is an illness caused by a microorganism, virus, or other disease-causing substance that can spread and cause serious illness or death.

Infectious Knowledge

On November 4, 1922, Englishman Howard Carter discovered an ancient doorway that led to the legendary tomb of Pharaoh Tutankhamen, the boy-king. The discovery was one of the most important of all time, and it created tremendous euphoria for many, especially the dig's benefactor, English Lord Carnarvon. Yet, for some, the discovery was not welcome. It was widely reported by the media that an ominous inscription in the tomb "Death shall come on swift wings to him that toucheth the tomb of Pharaoh" was a curse for all those who violated the King's tomb.

In early April 1923, six weeks after opening the burial chamber of Tutankhamen's tomb, Lord Carnarvon died from unusual complications resulting from a mosquito bite. Within six years, 11 people connected with the discovery of the Tomb had died of unnatural causes. The media, already propagating "The Mummy's Curse" followed each case with great scrutiny, and by 1935, had attributed 21 deaths to the "Curse."

Did the media invent "The Curse" or was there something real here? Recently, scientists identified several potentially dangerous mould spores following an examination of numerous mummies. It was suggested that when a tomb is first opened, a pressure gradient of air is formed allowing air to rush in which may disperse and aerosolize small spores within the tomb. Mould spores are extremely resilient and can survive thousands of years, especially in a dry tomb. Once airborne, spores can enter the body through the nose, mouth, or eye. Some moulds are particularly deadly and a relatively small dose of spores can cause serious disease or even death. For this reason, archaeologists now wear protective masks and gloves when unwrapping a mummy. The Mummy's Curse may have been nothing more than mould spores, or was it?

Some disease-causing organisms already live inside of us and others live in the environment. Some, like the bacterium that causes anthrax, can survive for a long time outside of a *host*. Others, like the agents that cause Lyme disease and malaria, have complex life cycles that involve ticks, mosquitoes, or even snails. In these cases, the mosquitoes or ticks that bite us and pass the organism along are called *vectors*.

Disease Diction

A **host** is a live organism (including humans) in which a disease-causing organism lives, multiplies, and causes disease.

A **vector** is a biological vehicle—often a free-living organism—that carries an infectious organism and passes it to an animal or human host.

A person doesn't necessarily have to be sick or show any signs of illness to be infected with, or even spread, an infectious disease. For example, the bacteria that causes the dangerous staph infection (discussed in Chapter 22), is commonly found in people's noses without causing any infection. Doctors, nurses, and health care workers who have the staph bacteria in their noses can serve as carriers of staph and shed bacteria onto surfaces or directly onto patients, causing serious infections even though they themselves are not sick.

Typhoid Mary is one of the most famous cases in medical history of someone who carried a disease but did not get sick from it herself.

Typhoid Mary: Disease Carrier Extraordinaire

Mary Mallon immigrated to the United States from Ireland in 1883. In the summer of 1906, she went to work as a cook for a rich New York banker, Charles Henry Warren. She worked for Mr. Warren and his family at a home they had rented in Oyster Bay, New York. At the end of August, 6 of the 11 people living in the house came down with typhoid fever.

Infectious Knowledge

Typhoid fever is caused by a bacterium that is spread through water or food. In the nineteenth century, typhoid fever, which causes headache, loss of energy, upset bowels, and a high fever, killed about 10 percent of those who got it. Once health officials learned that a clean water supply prevented most typhoid fever, the death rates from the disease began to fall.

The doctors who examined the sick members of the Warren household couldn't find contaminated water or food that would explain the outbreak of typhoid fever. The owners of the house, afraid they wouldn't be able to rent it again, hired a sanitary engineer to find the source. After striking out again with food and water sources, he began to investigate the people living and working in the house. He discovered that the family had changed cooks on August 4, three weeks before the typhoid outbreak. The new cook was Mary Mallon.

Mary had suffered a mild case of typhoid at some point in the past, and it turned out she was still a carrier, although she wasn't sick herself. Despite health officials' strong suspicion that Mary Mallon was the cause of the Warren-household typhoid outbreak, Mary refused to cooperate with the authorities and get tested. When officials investigated her work history, they found that seven of the eight families she had worked for during a 10-year period had suffered from typhoid outbreaks. In all, 22 people developed typhoid fever, and one had died.

Finally, in March 1907 the New York City health inspector had Mary tested against her wishes. Health officials' suspicions were confirmed: Tests showed that Mary was a typhoid *carrier*. She was then put into isolation on North Brother Island in the East River of New York City. In 1910, after three years, she was allowed to leave, on the condition that she stayed in touch with the health department and didn't work with food.

Upon release, Mary worked as a laundress, but she couldn't earn enough money to live, so she started cooking again. Health officials found her in 1915, using an alias and working as a cook in Sloane Maternity Hospital in New York City. During three months there, she had spread typhoid to at least 25 doctors, nurses, and staff. Two people had died. So Mary Mallon found herself back on North Brother Island, where she lived for 23 more years until her death.

When she died in 1938, there were 237 other typhoid carriers who had been identified and were living under city health department observation.

Disease Diction

A **carrier** is a person who has a disease-causing organism on their skin or in their body. Carriers often are not sick, but can spread disease if they come in contact with others.

The Ongoing Battle Against Disease

In the last century, we have made many advances against infectious diseases, and even eliminated a number of them. Smallpox has been completely eradicated from the planet (with the exception of a few samples of it kept for research purposes—see Chapter 6 for more on this). Polio, diphtheria, and a number of childhood diseases have been almost eradicated due to the development of effective vaccines (see Chapter 15).

The discovery of antibiotics, drugs that are effective at halting some bacterial diseases, also changed our approach to infectious diseases and opened the door to modern medicine. With the help of antibiotics, we found cures for diseases like tuberculosis, soldiers stopped dying from wound infections, post-surgical infections could be better

Disease Diction

An **acute** infection is a disease with severe symptoms that are generally of a short and self-limiting duration. The common cold is an example. A **chronic** infection is a disease that persists over a long period of time and often needs ongoing or periodic treatment. Hepatitis C is a chronic disease.

controlled, and bacterial pneumonias almost disappeared. The pre-1940s hospitals were dominated by *acute* infectious disease cases, while hospitals in the antibiotic era saw a shift toward *chronic* care cases like heart disease and cancer, because formerly deadly infectious diseases could be controlled or cured with antibiotics.

Yet, there have been some major warning signs that the war on infectious diseases is far from over. The emergence of new and devastating diseases like HIV, as well as the re-emergence of diseases like tuberculosis, especially in dangerous drug-resistant forms, reminds us that we still have a long way to go. The advancements that allow doctors to perform invasive medical procedures and keep sick people alive longer have also made us vulnerable to infections people get in hospitals.

Major Outbreaks: Epidemics and Pandemics

Disease Diction

An **outbreak** occurs when a group of people in a small geographic area become ill with an infectious disease. An **epidemic** occurs when an infectious disease spreads beyond a local population, lasts longer, and reaches people in a wider geographic area. A **pandemic** occurs when an epidemic spreads around the world.

Epidemics occur when an infectious disease spreads beyond a local population, lasting longer and reaching people in a wider geographic area. When that disease reaches worldwide proportions, it's considered a *pandemic*. A more localized occurrence of a specific infectious disease is called an *outbreak*.

How does an outbreak become an epidemic or pandemic? It depends in part on how easily the organism that causes the disease moves from person to person. It also depends on the behavior of individuals and societies. In this book, we will talk about a variety of epidemics and pandemics that have occurred throughout history.

Infectious Knowledge

One of the major pandemics of the twentieth century was the Spanish influenza epidemic of 1918. That flu killed nearly 40 million people worldwide in less than one year. Roughly 82,000 people died every day from the flu. By comparison, 9 million people died during the five years of World War I.

Of course, no force of weather, geology, war, or disease—not even AIDS—comes close to the horror of the bubonic plague, which lasted for years and decimated populations.

Public Health Sleuths: Epidemiologists

The study of epidemics is a branch of science called *epidemiology*. Epidemiologists are part doctors and part sleuths, using historical information, contact tracing (finding all people a person with a disease had contact with), medical records, interviews with patients and family members, computer mapping, genetic information about different strains of organisms that cause disease, and other methods to determine how and why diseases spread. Recent developments in the field of epidemiology, including advanced computers and molecular technology, are allowing epidemiologists to identify epidemics far earlier than in the past.

Cases of disease are reported two ways:

♦ **Incidence** is the number of new cases within a given time period. It shows how quickly a disease is spreading.

♦ **Prevalence** is the absolute number of cases in a given population—either at a point in time or over a period of time. Unlike incidence, prevalence includes both old and new cases, so it shows the impact of a disease on a population.

Disease Diction

Epidemiology is the study of epidemics. Where they happen, when they happen, what factors allow them to happen, and what can be done to contain and prevent them.

Infectious Knowledge

The Centers for Disease Control and Prevention (CDC) is a federal agency under the U.S. Department of Health and Human Services. It is the leading federal agency for protecting the health and safety of Americans at home and abroad, providing information to enhance health decisions and promoting health through strong partnerships.

The CDC is the national focal point for disease prevention and control, environmental health, and health promotion and education. Its mission is to promote health and quality of life by preventing and controlling disease, injury, and disability.

Specific areas of focus include monitoring health, detecting and investigating outbreaks of disease, conducting research to enhance disease prevention, developing and advocating sound public health policies, improve disease prevention, promoting healthy behavior, fostering a safe environment, and providing leadership and training.

The CDC builds partnerships with state and local health departments, corporations, and other private entities to accomplish its goals.

Advances in Public Health ...

The twentieth century brought incredible changes throughout the world, many for the better. One of the most far-reaching advances has been in the arena of public health. Thanks to improvements brought about due to research and programs instituted by public health officials, people now live an average of 30 years longer!

Major Public Health Challenges

Despite achievements, there are other diseases and health problems to handle. The increase in world travel has led to increased chances of acquiring rare or exotic diseases. The shrinking world also means that people from other countries arriving in the United States bring the risk of new and reemerging strains of old diseases and other health problems. Other challenges include ...

> **Antigen Alert**
>
> Infectious diseases account for one third of all deaths in people 65 years and older. Early detection is more difficult in the elderly because the typical signs and symptoms are frequently absent. A change in mental status or decline in function may be the only obvious problem in an older patient with an infection.

◆ The large number of antibacterial products on the market and their misuse adds to the ever-growing problem of drug-resistant bacteria.

◆ Gaps in adolescent immunization schedules, which allows for the possibility of the reemergence of epidemics of various childhood diseases.

◆ Alarming levels of risk-taking behavior, by adolescents and others, leading to the spread of sexually transmitted diseases.

◆ Large percentages of uninsured adolescents and adults, leading to delays in diagnosis and treatment of infectious diseases.

The Threat of New Diseases

Medical science has made tremendous strides in overcoming infectious diseases in the twentieth century. Despite this, several epidemics of previously unrecognized diseases have occurred during the last 20 years. These diseases include Lyme disease, Legionnaires' disease, *toxic shock syndrome*, West Nile, and AIDS. Examination of past epidemics, including the plague of Athens, the Black Death, syphilis, and influenza, suggests that the sudden occurrence of diseases that were previously unrecognized is not unusual.

Epidemics of infectious diseases may arise by several mechanisms, including changes in an organism that make it more potent or able to cause disease more easily, or the introduction of an infectious agent into a population that has never been exposed to it before and, therefore, has no natural immunity. Environmental and behavioral factors may play an important role, too, as illustrated by toxic shock syndrome and AIDS.

Disease Diction

Toxic shock syndrome is an illness caused by a toxin, or poison, secreted by the bacterium *Staphylococcus aureus*. The disease is very rare. It occurs mostly in menstruating women using high-absorbency tampons. These tampons can facilitate the infection because their prolonged use enhances bacterial growth. Symptoms include high fever, vomiting, diarrhea, rash, red eyes, dizziness, lightheadedness, muscle aches, and decreased blood pressure. More severe symptoms can be shock, kidney failure, and liver failure.

Toxic shock cases first appeared in the early 1980s and the majority of them were linked to the use of a particular high-absorbency tampon that was removed from the market.

To help prevent toxic shock syndrome, women should use the lowest-absorbency tampon they require, alternate tampon use with pads, and change tampons often.

There are a number of factors that lead to the emergence of new threats from infectious diseases. Experts generally agree on six of them:

- **Environmental changes and disturbances to the balance of natural habitats.** Many environmental factors contribute to the infectious disease threat. They range from things like global warming to the tremendous increase in the number of dams built in Africa over the past 20 years that has led to an increase in waterborne infectious diseases.

- **Human demographics and behavior.** Poverty, urbanization, and poor sanitation have been linked to increases in infectious diseases for hundreds, if not thousands, of years. Recently, the movement of refugees has also become a factor that allows diseases to spread. Sometimes these factors mean that diseases that we cure regularly in the United States still run rampant in other parts of the world.

- **International travel and commerce.** Throughout history, increases in trade and international travel have led to outbreaks and epidemics of infectious diseases. Advances in the twentieth century sped that process, particularly due to air travel. An infected person can get on a plane anywhere

Potent Fact

Having a cure isn't always enough. Without access to medical care, millions of people still suffer from curable diseases. Tuberculosis can be cured cheaply, but millions still die from it each year.

and, within 24 hours, carry a new disease to his or her destination. There have been documented cases of people with tuberculosis, for example, traveling on planes and infecting other passengers, as well as people who have gotten sick once they are in the United States with strains of tuberculosis that have been traced directly back to Russian prisons. Rapid global travel greatly increases our vulnerability.

◆ **Immune suppression due to acute and chronic illness.** A weakened immune system makes us more vulnerable to certain diseases. Sometimes it is a temporary problem, like infections that cancer chemotherapy patients may get that go away once their treatment is completed. Other times, like with HIV, patients will be extra-vulnerable throughout their lives because the disease permanently weakens the immune system.

◆ **Microbial adaptation and change.** Microbes, the tiny organisms that cause disease, have been around for 3.5 billion years. During that time, they have developed a variety of methods to withstand the environment, including the human immune system, so that they can survive. It is only in the past 60 years or so that we have developed drugs that can kill or debilitate disease-causing microbes. But it didn't take long for the microbes to change in ways that allowed them to defeat those drugs and continue to survive and multiply. It is a constant battle for survival, and if we don't continue to develop new ways to outsmart the microbes, they will certainly continue to outsmart us.

◆ **Breakdown of public health measures.** Despite a century of tremendous advances in public health, we still have a long way to go in some areas. For example, we have excellent vaccination programs in this country that have cut down significantly on infectious diseases that afflict children. However, if we are not vigilant about vaccination, these diseases could come back, as they have in other parts of the world where vaccination programs are not as thorough as they are here.

Catching On to the Cycles

Fortunately, even without effective drugs and other public health measures, epidemic diseases tend to slow down over time because of changes in the infecting organisms and changes in the host. Study of past and current epidemics reveals that they are cyclical: New diseases will arise periodically, occasionally with a devastating outcome. With time, the effects of these diseases on the population will go away. The cycle will begin again when a new disease emerges.

The Least You Need to Know

- Infectious diseases are caused by microorganisms that can be passed from one person to another.

- Epidemics and pandemics occur when infectious diseases spread rapidly through a broad geographic area. Studying and understanding them can help prevent them from recurring.

- Despite optimism about their eradication, infectious diseases are still responsible for one-third of all deaths worldwide each year.

- There are a number of factors that contribute to our continuing vulnerability. Some of these factors are man-made and some are natural.

A History of Infectious Diseases

In This Chapter

- ◆ How early epidemics savaged millions of people
- ◆ First attempts at fighting off disease
- ◆ Bacteria and viruses are identified
- ◆ Coming to terms with drug resistance and new epidemics

You might be surprised to learn that our long-ago ancestors, people who lived in small groups and traveled around in search of food, didn't have as many problems with infectious disease as we do today. Infections, after all, do the most damage when they spread to large populations. When people lived in small groups far apart from other groups, they didn't provide the right environment for infections of epidemic proportions.

In this chapter, you'll learn about the evolution and history of infectious diseases, early discoveries and identifications of their causes, the development of treatments, and the continuing persistence of infectious diseases despite our medical advances to contain and control them.

Disease and Development Go Hand in Hand

Our early ancestors lived in small groups, moved often in search of food, and didn't have domesticated animals. Their survival was determined by the availability of resources like food and water. The infections they got were most likely from trauma and accidents or from eating diseased animals.

As the population of large animals to hunt decreased, people began to rely on domesticated animals and cultivated plants for food. They settled in villages in order to grow their food and raise animals. Once people started living in close proximity, they became easy targets for diseases, which were successful because they quickly spread from one person to another. Over time, survival became more closely related to disease than to the availability of food and water.

Villages grew into towns and towns grew into cities. As people moved closer and closer to each other, they provided more opportunities for diseases to spread. Often people and animals lived under the same roof, which hastened the spread of disease. Sanitary conditions in highly populated areas were often very poor, with people living in close proximity to exposed garbage and raw sewage, creating favorable conditions for the spread of many contagious diseases. People were exposed to influenza, salmonella, tuberculosis, and worms. Some spread through the air; others came from contaminated water or food. Insects were also a danger because of the disease-causing organisms many of them carried and spread.

> ### Infectious Knowledge
>
> Just because our early hunter-gatherer predecessors didn't have as many diseases to worry about as we do, it doesn't mean that they lived long, healthy lives. It's estimated that the average lifespan of an early hunter-gatherer was less than 40 years, whereas today the average lifespan is 77 years.

> ### Disease Diction
>
> **Immunity** occurs after a person has been exposed to a particular disease. For example, if you've had chickenpox as a child, your body responded to the infection and produced a multi-level biochemical response. Part of that response was to create cells that fight chickenpox, as well as memory cells that can reactivate the fighter cells quickly if the body sees the same invader again. See Chapter 4 for more about immunity.

Epidemics: The Downside of Travel

As cities developed, people began to travel. Traders and armies moved around, spreading disease with them. As they went to new places, they were often carriers of diseases that didn't make them sick but caused others to get sick. Why? Because those who had been exposed to a disease before had developed *immunity* to it. Those who hadn't been exposed had no immunity and therefore were vulnerable.

In addition, some of these travelers brought animals with them and the animals carried diseases as well.

Diseases that were new to a population often spread like a case of chicken pox in a room full of kindergartners. As a result, epidemics began to occur. Parts of the world like the Middle East and India were used to traders and travelers, so people living in these regions had been exposed to more disease-causing organisms and therefore had a chance to develop immunity to them; as a result, they were spared from the worst epidemics. Greece, Italy, and China weren't so lucky:

♦ In 430 B.C.E., the plague of Athens occurred when, during war with Sparta, 200,000 villagers left the countryside and came to the city. Crowded together in the city, the Athenians began getting sick with an unidentified infection that was spreading throughout the eastern Mediterranean. Preventive measures failed and people died in large numbers. It is estimated that one quarter to one third of the population died.

♦ In 160 C.E., the bubonic plague led to the collapse of the Han Empire in China. (For more on the bubonic plague, see Chapter 6.)

♦ In 166 C.E., Roman troops returning from Syria brought a plague that devastated the Roman Empire and killed between four and seven million people in Europe. The resulting social and political upheaval helped lead to the collapse of the Roman Empire.

♦ The bubonic plague epidemic that lasted from 1346 to 1350 moved from China along trade routes to Russia and throughout Europe. One third of the European population was killed.

♦ In the sixteenth century, the Spanish introduced European and African infections into Central and South America. Within 10 years, Mexico lost one third of its population. Over a 75-year period, 95 percent of Mexicans died. (This was how the Incas and the Aztecs were conquered.)

Infectious Knowledge

How did Cortez conquer the Aztecs? How did Pizarro overthrow the Inca? Cortez and Pizzaro had better weapons, and they had horses. But they were outnumbered. Pizarro had 168 soldiers. The Inca leader Atahuallpa had 80,000! There is no way the Conquistadors could have overthrown these civilizations unless they had a secret weapon. They did, even though they didn't know it. And they certainly didn't know how powerful it was. The weapon was smallpox. The disease traveled ahead of them, weakening the Inca and Aztec so they couldn't put up a good fight. Many Inca were dead before Pizarro even showed up.

Fighting Infection: The Early Years

Although no one knew what caused the diseases that were killing millions of people throughout the world, people began to learn ways to keep them from spreading. These early techniques weren't usually successful, but people were no longer letting diseases ravage their villages, towns, and even whole countries without putting up a fight.

Quarantine and Enforced Isolation

One of the most effective early methods of fighting infection was quarantine or isolation. This included isolation of the sick, restriction of movement for the sick *and* the healthy during disease outbreaks, and isolation of healthy people.

A quarantine sign indicating the presence of an infectious disease. Quarantine has been an effective means of preventing the spread of disease for centuries.

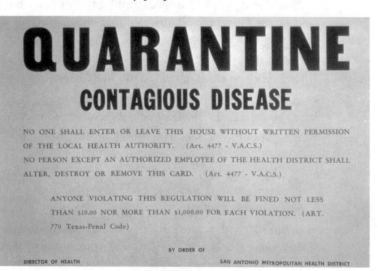

QUARANTINE

CONTAGIOUS DISEASE

NO ONE SHALL ENTER OR LEAVE THIS HOUSE WITHOUT WRITTEN PERMISSION OF THE LOCAL HEALTH AUTHORITY. (Art. 4477 - V.A.C.S.)
NO PERSON EXCEPT AN AUTHORIZED EMPLOYEE OF THE HEALTH DISTRICT SHALL ALTER, DESTROY OR REMOVE THIS CARD. (Art. 4477 - V.A.C.S.)

ANYONE VIOLATING THIS REGULATION WILL BE FINED NOT LESS THAN $10.00 NOR MORE THAN $1,000.00 FOR EACH VIOLATION. (ART. 770 Texas-Penal Code)

BY ORDER OF

DIRECTOR OF HEALTH   SAN ANTONIO METROPOLITAN HEALTH DISTRICT

> **Potent Fact**
>
> Leprosy is a chronic infectious disease that attacks the skin, peripheral nerves, and mucous membranes. It is most common in warm, wet climates. Leprosy's victims have been despised throughout history and kept separate in leper colonies and sanitariums. Even today people with the disease are shunned and ostracized. In fact the contemporary meaning of the word leper, "one to be shunned or ostracized" is derived from early attempts, through quarantine, to halt the spread of disease.

In biblical times, lepers were isolated from the rest of the population, both to avoid the spread of this mutilating disease and so people didn't have to see its horrifying effects. In the fourteenth century, parts of Italy and Russia banned travel to try to control the bubonic plague. They hoped to prevent healthy people from getting sick. Though not always effective, quarantine did help. Sometimes it still does.

Primitive Treatments

Through careful observation and trial and error (combined with a hefty dose of good luck!) people sometimes even unwittingly hit upon a cure. Jesuit missionaries in malaria-ridden Peru had noted the native Indians' use of Cinchona tree bark to treat the disease. In 1627, the Jesuits imported the bark to Europe for treating malaria. The bark contained quinine, which as you'll discover in Chapter 9 is still used to treat malaria today.

And there is evidence that even our very ancient ancestors may have known how to treat some infections: Researchers have discovered that the mummy called "Ice Man," who lived approximately 5,000 years ago, was infected with a parasitic worm. Interestingly, scientists also discovered that the ancient man had ingested a tree fungus that contained oils capable of killing the worm!

Early Vaccination Method?

The Asian emperor Mithridates even habituated himself to what would normally be lethal doses of poisons by gradually increasing the amount he ingested. He was unknowingly building up and training his own immune system. An early, albeit not recommended, immunization technique.

Cleaning Up the Water Worked, Too!

In 1854 British physician John Snow showed a link between sewer-tainted drinking water and the spread of cholera, a common waterborne disease that causes severe vomiting and diarrhea. (See Chapter 13 for more on cholera.) The measures that were taken in response to this discovery, including improving sanitation and providing a safe water supply, helped to increase life expectancy.

Fortunately, many ancient populations had clean water and good waste disposal. This was partly to appease the gods they believed in, as they thought that the gods punished people by inflicting them with diseases, and it was partly thought of as necessary for personal health. Ancient Rome had a

> **Infectious Knowledge**
>
> British physician and epidemiologist Dr. John Snow, 1813–1858, was one of the pioneers in developing links between science and health. His work helped to shape modern medicine and public health practices. His use of epidemiological techniques, like mapping, to isolate cholera outbreaks and identify contaminated water pumps, was revolutionary.

sophisticated network of sewers and many of the wealthy had latrines. If not for these early attempts at good hygiene and sanitation, many more people would probably have died.

> **Potent Fact**
>
> In 1847, Ignaz Semmelweiss showed that patients of doctors who did autopsies suffered from more infections than patients of midwives who didn't do autopsies. His hypothesis was that infection was spread by doctors' contaminated hands. He recommended hand-washing with chlorine. He showed that this practice lowered mortality rates significantly. Despite his results, his work was largely ignored at the time, though today proper hand-washing is known to be vital in stopping or slowing the spread of infection.

Early Theories on Causes of Infection

As noted previously, in ancient times many people thought that disease was sent down from the gods as punishment, although some people believed that poisons and vapors that came from the decay of living materials played a role.

In 1546, Italian physician Girolomo Fracastero showed that infections, such as rabies, are produced by "seeds of disease." Later that same century in 1589, Thomas Moffet suggested that lice, fleas, and mice could cause disease. It wasn't until the 1840s, however, that Jacob Henle said that disease was caused by living particles that acted as parasites in people. He began to define a disease-causing organism: It was present when someone was sick, it could be isolated, and it could cause disease in a healthy person.

In 1683, Anton van Leeuwenhoek used a microscope to see bacteria that he had scraped off of his own teeth. That led the way to visualizing some of the dreaded causes of infectious diseases.

> **Disease Diction**
>
> A **parasite** is an organism that grows, feeds, and is sheltered on or in a different organism, such as an animal or a person, while contributing nothing to the survival of its host. See Chapter 3 for more on parasites.

> **Infectious Knowledge**
>
> Pasteur also developed a veterinary vaccine for anthrax and one for rabies.

Germ Theory: The Real Story

Louis Pasteur helped reveal the vastness of the microbial world and its many practical applications. In 1857, he found that microbes, also called microorganisms, were behind the fermentation of sugar into alcohol and the souring of milk. He developed pasteurization, a heat treatment that killed microorganisms in milk. Once pasteurized, milk no longer spread diseases such as salmonella. A major break-through had been made.

In 1864, Pasteur stated that infectious diseases were caused by living organisms. He called these organisms germs.

Shortly thereafter, Robert Koch isolated the bacterium that causes anthrax in horses. He showed the bacteria's ability to form spores and cause disease. He also showed that a specific organism causes a specific disease. Koch developed four rules that could be used to confirm the link between a specific organism and a specific disease. They are called Koch's Postulates, and they state ...

- ◆ The organism must be present in every case of the disease.
- ◆ The organism must not be present in any other disease.
- ◆ It must be possible to isolate the organism.
- ◆ After growth in a culture, the organism must be able to produce disease.

Although we now know that some organisms can cause more than one disease, for the most part these assumptions still hold true today and have formed the foundation of all our ideas about how to identify the cause of an infectious disease.

In 1892, the Russian microbiologist Dmitri Ivanowski discovered tiny infectious agents that could pass through filters that stopped bacteria. Ivanowski called these microorganisms filterable viruses.

Disease Diction

A **microorganism** is a living thing, such as a bacteria or fungi, small enough so that it can be seen only under a microscope and not with the naked eye.

Potent Fact

Koch is also credited with discovering the bacteria that causes tuberculosis.

Infectious Knowledge

A lot of people confuse bacteria and viruses. **Bacteria** are organisms that can survive on their own. They have a cell wall that can be stained for identification. **Viruses** are smaller and can't live on their own. They are parasitic and must live inside the cells of a host. After entry into a host cell, viruses use the machinery of that cell to reproduce. See Chapter 3 for more on these infectious agents.

The Search for Cures

Throughout history, people have been trying to figure out ways to cure and prevent disease. Remarkably, some things that were done thousands of years ago were effective then and continue to be effective today. At the same time, we are constantly challenged by new diseases like AIDS, which we have not yet been able to cure.

Vaccination

One of the best ways to fight disease is to keep people from getting it in the first place. Vaccination is one method of accomplishing that. Vaccination depends on building a person's immunity to a particular disease-causing agent by exposing them to an avirulent (nondisease-causing) form of it without making them sick. There are two ways to do this:

◆ **Passive immunity** is provided by sources other than the person with an infection. It can be transferred by giving healthy people secondhand immune agents. For example, an animal that was previously infected with a disease likely will have developed antibodies, or immune agents, in their blood that fight off the disease. When these antibodies are taken from the blood of the animal and injected into the blood of humans or other animals, these now-"secondhand" immune agents can help fight off the disease in someone who has never before come in contact with the germs that cause it. This method has been used to provide immunity from diphtheria.

◆ **Active immunity** can be created by injecting weakened poisons or other proteins of an infectious organism into healthy people, thereby stimulating their immune system to produce a multi-faceted protective response. This method has been used to provide immunity from tetanus. Active immunity can also be produced by injecting attenuated, or weakened, cultures of disease-causing organisms into the body. Pasteur was the first to demonstrate this phenomenon with rabies and anthrax. The polio vaccine is also attenuated.

Infectious Knowledge

For hundreds of years in Asia, smallpox scabs of infected patients were dried, aged, and then blown into the nostrils of the uninfected. This produced mild attacks and built immunity. In Europe, the same principle was used when they removed pus from infected lesions and scratched it into the skin of healthy people.

In 1796, Edward Jenner observed that milkmaids were not getting smallpox. His research uncovered the fact that they were immune because of exposure to cowpox, a virus that is related to smallpox. By scratching cowpox into the skin of healthy people, smallpox was prevented. One hundred thousand people were "vaccinated" by 1801.

Vaccines Today

Today we have vaccines for influenza, tetanus, measles, polio, smallpox, and a variety of other diseases. Vaccines are the best way to combat viral infections, but they are often difficult to make and test.

Vaccination is not always successful and it can't protect everyone. It is a particularly weak approach when there are many different strains of an organism causing disease, like influenza, or when the organism changes rapidly, like HIV, or against new diseases caused by unknown organisms.

The Discovery of Antibiotics—Miracle Drugs of the Twentieth Century

Anti-infective agents have been used for thousands of years. Twenty-five hundred years ago, the Chinese used bean curd to treat skin infections. Sulfur was used to treat scabies, and mercury was used to treat syphilis.

In the late nineteenth century, it was found that fungi and yeasts could destroy bacteria. One organism was harmful to another. This was called antibiosis, the basis for the word *antibiotic*. Other researchers began to discover and classify additional antibiotics and their effects. Today, there are more than 25,000 antibiotics that fall into one of six major classes. The classes refer to the drugs' mechanisms of action—the way they act to kill or disable infectious bacterial invaders.

In 1928, Alexander Fleming discovered the antibacterial activity of a fungus called *Penicillium notatum*. By 1940, penicillin was shown to protect mice against a number of different bacterial infections. Penicillin was produced to use in World War II and was in general use by the end of the 1940s. It was considered a miracle drug, although some infections were not cured by it—an early indicator that some bugs were able to become resistant to it (more on resistance later in this chapter).

Many other antibiotics have been discovered, manufactured, and sold since the introduction of penicillin. In addition to those that occur naturally, drug companies have been able to make synthetic versions of some drugs. The discovery of the first antibiotics and the explosion in drugs that came afterwards created great optimism about our ability to conquer infectious disease.

Disease Diction

An **antibiotic** is a chemical that is produced naturally or can be manmade, that either prevents the growth of microorganisms or kills them.

Potent Fact

Some antibiotics are bacteriocidal. This means that they kill the bacterial invaders. Others are bacteriostatic. They prevent new growth and cell division of the invader. Bacteriostatic drugs require more work from the immune system to help clear an infection.

Other Twentieth-Century Advances

Other changes and advances that occurred in the twentieth century also were cause for optimism.

The Molecular Revolution

The pivotal discovery that genetic information resides in DNA arose from studies on pneumococcus, the bacteria that causes pneumonia. These studies were done to monitor the epidemic spread of pneumonia that was occurring in the 1920s. In 1928, British physician Frederick Griffith found that extracts from a disease-causing strain of pneumococcus could change a harmless strain into one that caused disease. In 1944, Oswald Avery, Colin MacLeod, and Maclyn McCarty reported that DNA was the transforming factor, work that earned a Nobel Prize.

Disease Diction

Molecular genetics is the study of the flow and regulation of information related to heredity between DNA, RNA, and protein molecules.

DNA, which stands for deoxyribonucleic acid, is the building block of life. DNA is the molecule that encodes genetic information. It's what determines our hair color, for example. DNA codes for 20+ amino acids, which then combine to form proteins. DNA is made up of four nucleotides or nitrogeneous bases. The sequence of these nucleotides or base pairs determines the genetic make-up of each organism.

Those findings prompted interest in the life cycle of bacteria and the construction of chromosome maps, or outlines of their genetic make-up, for them. The science of bacteriology took center stage, and many developments in molecular genetics and biotechnology that followed came out of research done in the mid-twentieth century.

The Century of Optimism

The twentieth century saw great strides in public health. In the United States, the average life span lengthened dramatically from 47 years in 1900 to 77 years by 2001. Similar trends occurred in other countries, although the gains have been smaller in economically and socially depressed countries. Deaths from infectious diseases declined as well.

Childhood immunization played a major role. So have public health measures like the protection of food and water supplies, segregation of coughing patients, and improved personal hygiene. Overall economic growth contributed to less-crowded housing, improved working conditions, and better nutrition.

Healthier lifestyles, including less smoking, better diets, more exercise, and better hygiene and food preparation have been important as well. Preventive medications and medical and surgical interventions have also kept people alive longer.

Potent Fact

Public health experts say that infectious diseases are the single leading cause of death worldwide.

The 1940s and 1950s were notable for wonder drugs—antibiotics like penicillin and strep-tomycin and a growing list of others that promised an end to bacterial disease. Vaccines were developed to combat some viral diseases—polio and smallpox, for example. Through a massive public health campaign, smallpox was eradicated around the globe.

Confidence about medicine's ability to fight infectious disease was so high that by the late 1960s some felt infectious diseases were largely defeated. They suggested that researchers should shift their attention to heart disease, cancer, and psychiatric disorders. These views caused shifts in the priorities for research funding and drug development.

The optimism and complacency of the 1960s and 1970s was shattered in 1981 with the recognition of HIV, which has become a modern-day global epidemic with terrible consequences. Since then, we have also seen a frightening number of diseases that are now resistant to antibiotics.

Antibiotic Resistance: A Warning Among the Advances

The development of antibiotic resistance is a matter of great concern. As noted previously, resistance to penicillin appeared before 1950. For a while there were so many new drugs being discovered and manufactured that there was little worry about resistance. Unfortunately, that is no longer true, and we are reaching a point where some infections—especially those spread in hospitals, where the patients' immune systems are already weak—are difficult, if not impossible, to treat. If not addressed with appropriate steps, the spread of resistance may put many of our recent medical advances at risk.

How Does Resistance Happen?

If bacteria were people, they'd be the sly, criminal element of the population, constantly changing their guises and tactics in order to evade capture and prosecution. These bad bacteriological bugs have developed several ways to resist the antibiotics developed to "arrest" them:

- ◆ They can keep the drug from reaching the target by making the outer wall of their cell impermeable so the drug can't get in.

◆ Like miniature bilge pumps, they can pump drugs out of their cells, thereby preventing them from having any effect on the disease-causing bacteria.

◆ They can produce enzymes that inactivate or destroy the drug.

◆ They can alter the drug target so the drug no longer attaches to the bacterial component it is meant to act on.

If resistance develops through genetic mutation, it is passed on to subsequent generations as the bugs multiply. Resistance can also be acquired by the transfer of genetic material from one organism to another. When the new organism multiplies, the new gene can be incorporated into the DNA and passed on to future generations. Resistant strains are particularly problematic when they spread in a hospital where many patients have temporarily or permanently weakened immune systems.

Why Does Resistance Happen?

Resistance can be caused by overexposure or misuse of antibiotics. For example, many people want prescription medicine when they have colds. Unfortunately, antibiotics don't work on colds because colds are caused by viruses, and antibiotics only kill bacteria. However, many doctors prescribe antibiotics anyway. The more an organism "sees" an antibiotic, the greater the chances that organism will become resistant to that antibiotic. The kind of exposure that happens when an antibiotic is used where it won't work inadvertently increases the odds of resistant infections in the future.

Resistance can also be caused by inappropriate use or overuse of antibiotics. For example, a person has a bacterial infection and their doctor prescribes antibiotics for 10 days. After five days, the person feels better and stops taking the medicine. Not all the bacteria that caused the infection have been killed, so the ones that survive—the stronger ones, since the weaker are killed first—will thrive and multiply. When the patient gets sick again, these organisms will most likely be resistant to the drug the doctor prescribed the first time.

Infectious Knowledge

The annual cost for treating people who have infectious diseases in the United States is approximately $120 billion dollars.

Resistant infections can be passed from one person to the next, potentially causing epidemics of infections that are nearly impossible to treat. This is called horizontal transmission.

Outlook for the Future

Clearly, infectious diseases are still a challenge. The optimism of the early antibiotic era has diminished. We have seen new diseases emerge, and we have seen old diseases re-emerge—sometimes in deadly new drug-resistant forms. Unfortunately, with the speed of global travel today, a dangerous infection can spread worldwide within hours.

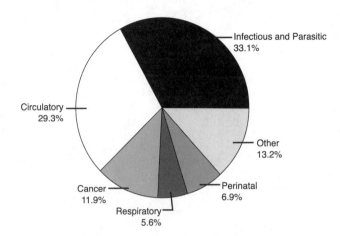

WHO chart showing world-wide mortality trends. Infectious diseases, the leading cause of mortality, account for one third of all deaths.

After the development of the germ theory of disease, diagnosis, treatment, and prevention took center stage. Since the 1950s, new diseases and antibiotic resistance have reminded us that the bugs are difficult to beat. To paraphrase Nobel Laureate Joshua Lederberg: The bugs have been around for 3.5 billion years. By comparison, we've been here for a much shorter time. They will continue to survive and thrive, and we must learn to live as peacefully with them as possible.

We must come up with new ways to use existing drugs, use new combinations of drugs, develop new drugs, and educate patients and doctors about the dangers of the misuse and overuse of antibiotics. The next time your doctor tells you that you don't need an antibiotic, listen. If he or she prescribes one, follow instructions and complete your course of treatment! You will be doing yourself—and the rest of us—a big favor.

The Least You Need to Know

- Infectious diseases have been around for a long time, long before people knew what caused them and how to cure them.
- Some early treatments for diseases worked even though people didn't know why.
- Antibiotics cure many infections, but antibiotic resistance continues to lessen their effectiveness.
- Vaccines are a proven way to combat infectious diseases, but they are difficult to develop and test.
- We must continue to develop new methods to attack infectious diseases.

Chapter

The Nature of Infectious Organisms

In This Chapter

- ◆ The basic biology of infectious organisms
- ◆ The difference between free-living organisms that cause disease and those that are parasitic and depend on their host for survival
- ◆ How microorganisms cause infections

Infections don't simply appear out of thin air—they are caused by something, albeit usually something too small for the human eye to see. The "somethings" that cause infectious diseases in people can be broken down into two general groups: those that can survive on their own, including bacteria and fungi, and parasitic organisms that need a host to survive. These include viruses and worms. In recent years scientists have also fingered a kind of protein called prions as disease-causers, too, and they make up their own unique category.

In this chapter, you'll learn about the microbial underworld: the miniscule organisms that attack the human body and cause disease. We'll first look at the organisms that can survive in the environment on their own, and then turn to those that need a host to survive.

Bacteria: 3.5 Billion Years Old and Still Kicking

Bacteria are an ancient form of life, having survived on earth for billions of years. Bacteria are so hardy mainly because of their structure. They are small cells, found in the environment either individually or clumped together. Unlike parasitic disease-causing entities, which will be discussed later in this chapter, bacteria have all the necessary cellular machinery to live on their own.

> **Disease Diction**
>
> **Infection** occurs when an organism enters the body, increases in number, and causes damage to the body in the process.

The most complex part of a bacterial cell is usually its surface, or cell wall. Some bacteria secrete chemicals that create a hard capsule on their outer surface and some may have flagella, which are extra appendages like arms and legs that allow them to move around. Others may have a variety of external projections useful to help them stick to their chosen habitat.

Bacteria.

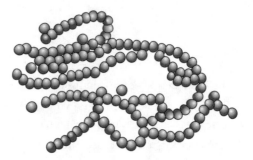

When bacteria multiply, the DNA, or genetic material, in the nucleus of a bacterial cell makes a copy of itself, and one cell divides to make two. The divided cells have the same genetic make-up as the original cell.

> **Disease Diction**
>
> **Commensalism** means "eating at the same table"—a neutral situation where the host and the bacteria live together, but have no effect on each other's life cycles. This is the relationship people have with most bacteria.

Most bacteria are fairly flexible about the conditions they need for growth. With the right temperature and a few simple nutrients, they are happy and will thrive.

How Bacteria Cause Infections

Bacteria are like people—their goal is to survive and prosper. Most bacteria we come in contact with—living either on us or in us—live in peaceful equilibrium with us, in what is called *commensalism*. But sometimes that balance is disturbed, and infection can result.

Coming in Contact

In order to cause infection, bacteria have to be able to survive in the environment and must "know" how to react when they come in contact with a potential host. This can happen in a number of ways and at a number of different speeds. Some bacteria cause disease quickly after finding a host. Others, like tuberculosis, can live dormant inside a host for years, causing disease when conditions become favorable.

Finding a Victim

Bacteria and other disease-causing organisms have developed many ways to move from the environment to a susceptible host, surviving long enough to cause disease. Some bacteria, like TB and anthrax, produce spores. Spores are outer coatings that are resistant to the effect of drying in the environment. Organisms that form spores can live for very long periods of time and then reactivate and cause disease when environmental conditions are favorable.

Some other bacteria, like the one that causes hospital staph infections, are able to survive on skin surfaces such as the hands and the nose—two important sources of transmission. Others, like the bacterium that causes cholera, survive in fluids like contaminated water supplies. Some, like the bacterium that causes salmonella, thrive in conditions of poor hygiene and are transmitted by fecal-oral spread, while others may require intimate contact to spread.

> **Infectious Knowledge**
>
> Anthrax is one of the most dangerous spore-forming bacteria. Once spores are formed, anthrax can live for years and years in hostile environments like the soil, reactivating and becoming virulent when conditions become favorable.

Disease Diction _____

Nosocomial infections are infections that are transmitted in hospitals. Some are opportunistic and cause disease because hospital patients are sick and weak. Others, like staph, may occur because of the nature of the hospital environment. The time, course, and severity of a given infection depend on the balance between the infecting agent's ability to cause disease (virulence) and the immune system's ability to fight it.

Making It Stick

Many body surfaces are washed by fluids. If a bacterium is going to multiply and cause an infection, it must have a way of physically "hanging on" and thriving in spite of being hit by liquid. To do this, bacteria make proteins that help them attach to host molecules, called receptors. They may also secrete sticky substances onto their surface to increase their chances of sticking.

Gaining Entry into Cells and Tissues

Sticking can be the first step to entering a host cell. Some bacteria do this. All viruses, which will be discussed later in this chapter, do it.

Entry into cells may lead to an infection that is limited to that cell type, or it may be a first step toward distribution of the bacteria throughout the body. Some bacteria are able to squeeze through spaces between adjacent skin cells and reach deeper tissues.

The Infection Cycle

Once bacteria have entered the body, there are a number of ways they cause disease. As long as they are able to evade the immune system, they may cause damage to local tissues. If they secrete poisons called toxins, these may travel through the bloodstream and cause damage to tissue and organs far away from the original site of the infection.

Evading Detection

The human body isn't going to let the bad bugs take it over without a fight. Our many defenses against infection make it a hostile environment for disease-causing bacteria. In order to survive, bacteria develop strategies to maintain their sources of nutrition. For instance, the body tightly controls its distribution of iron, a nutrient bacteria need to flourish, so most disease-causing organisms secrete proteins that "steal" iron from host proteins.

Choosing a New Victim

Once an infection has taken hold in the body, the next step is for the invader to find a new host to infect, usually in a way similar to how the present host got the disease. Bacteria that cause sexually transmitted diseases, for instance, produce a discharge to maximize their chances of being passed on (see Chapter 10 for more on this). Others produce diarrhea. Respiratory bacteria like TB produce nasal discharge, coughing, and sneezing to increase the chances that they can be passed through the air (see Chapter 8 for more on this).

Fungal Infections

About 180 of the 250,000 known species of fungi are able to cause disease in people. Most of these are moulds, but some are pathogenic yeasts. Some fungi are highly pathogenic and can establish a systemic infection in an exposed person. Other fungi are opportunistic—they live normally in or around us and only make us sick if our immune systems are weak.

These organisms are incredibly diverse and the infections that they cause vary greatly. Unlike bacteria, fungi have a true nucleus and rigid cell walls that make them immobile. Moulds are composed of numerous, microscopic branches known collectively as mycelium. During asexual reproduction, mycelia can form environmentally tough spores that are the principal means of transmission. Yeasts are small round cells, they exist singly and they reproduce by budding. In budding, the parent cell forms a bud, or outgrowth, on its outer surface as the nucleus divides. One nucleus migrates to the elongated bud. The cell wall material forms between the bud and the parent cell and the bud breaks away.

Disease Diction

An **opportunistic infection** is caused by an organism that normally lives at peace with us within our body or in the environment, but takes advantage of the opportunity a weakened immune system gives it to cause disease. Sometimes the opportunity is caused by another disease, like HIV. Other times the opportunity is a temporary suppression of the immune system, such as when a person is receiving chemotherapy for cancer.

Fungi.

The incidence of infections caused by fungi has increased dramatically in the past twenty years. These opportunistic infections are related to the growing population of people with weakened immune systems due to HIV, cancer, and other diseases; and to modern medical practices such as the invasive use of chemotherapy and drugs that suppress the immune system.

Parasites: Organisms That Cannot Survive on Their Own

Parasitic infections affect millions of people worldwide and cause considerable human suffering and economic hardship. Far from declining, many parasitic infections are increasing worldwide. Changes in climate induced by global warming have helped many parasitic diseases spread, while starvation and the breakdown in sanitation that accompanies war has led to the reemergence of others. Drug-resistance has also dramatically reduced our ability to treat and control many parasitic diseases.

Viruses: Tiny but Dangerous

Viruses are parasites. They cannot replicate outside a host cell and they cannot move on their own. These tiny organisms, which are much smaller than bacteria, are carried in water, food, wind, and blood and other bodily secretions. Once they invade a host, they can multiply rapidly.

Viruses contain either DNA or RNA, but not both. Because viruses need both DNA and RNA in order to reproduce, they invade host cells and take over their genetic machinery. When the host cell divides, the new cells also contain the virus.

Infectious Knowledge

All the genetic information living cells need to reproduce is carried in their DNA. When cells reproduce, DNA is copied into messenger RNA molecules that are used to make all the proteins needed to "build" a new cell. When cells divide, they make a complete copy of their DNA and end up with two identical copies of the cell's genetic information. The two daughter cells formed by cell division each carry exactly the same DNA present in the original cell.

Viruses don't have the DNA needed to make the thousands of proteins it takes to make even the simplest living cell. Instead, they carry a very small amount of DNA and RNA they use to make a few proteins, which then use the host cell and force it to make thousands of new viruses.

Viruses have no cell wall and no nucleus. They can't carry out functions that regular cells can, but they can take over a host cell and cause disease.

After bacteria were identified as causes of disease in the nineteenth century, it became clear that they weren't the only causes. Because viruses are so much smaller than bacteria, it was hard to isolate them. It wasn't until 1935 that Wendall Stanley found the virus that causes tobacco mosaic disease, an affliction of tobacco and other plants.

Virus.

The invention of the electron microscope has helped researchers identify many viruses. Because viruses multiply so quickly, they also evolve quickly and are difficult to treat and control.

The Viral "Life Cycle"

There are five steps in the replication cycle of all viruses:

1. **Attachment**   This is the stage where a virus comes into contact with a host cell and attaches to that cell on specific spots called receptors.

2. **Entry**   Once attached to a host cell, the virus secretes chemicals that weaken the host cell wall. Once there is a weak spot in the cell wall, the virus injects its DNA or RNA into the cell through that spot.

3. **Replication**   Once inside, the virus takes over the machinery of the host cell, instructing it to make more viral DNA and viral proteins.

4. **Assembly**   All cellular activity of the host cell helps to assemble new viruses. The host cell ends up filled with new viruses.

5. **Release**   After assembly, the virus particles secrete an enzyme that digests the host cell wall from the inside out. The new virus particles are released and they can infect new cells, starting the process over again.

Antigen Alert

Viral infections cannot be treated with antibiotics. If the doctor tells you that you have a cold, which is caused by a virus, don't try to get an antibiotic prescription.

Infectious Knowledge

The human immunodeficiency virus (HIV), which causes AIDS, is not what causes those it infects to die. The HIV virus attacks the immune system's T cells—the cells responsible for killing invaders—and kills them. The name "acquired immune deficiency syndrome" refers to the permanent weakened immunity caused by the viral killing of T cells. This leaves people vulnerable to opportunistic infections—infections caused by usually harmless organisms that take advantage of the opportunity provided by the lack of T cells. It is those other diseases that are the actual causes of death.

Viruses cause a wide variety of diseases, from the annoying but minor common cold to serious diseases like AIDS. Other viral diseases include polio, warts, hepatitis, smallpox, rabies, and influenza.

Protozoa: Single-Celled Organisms That Cause Multiple Diseases

Like viruses, protozoa must invade suitable hosts in order to complete their life cycles. They, too, are parasites.

Protozoa means "first animal." It refers to simple, single-celled organisms like amoebae. Some protozoa can form a protective cyst stage that helps them withstand harsh environmental conditions. Protozoa are classified into four different groups based on the way they move. Some protozoa move by pushing out the material inside the cells and then almost slithering along; others have hair-like extensions called cilia. Still others have flagella, which look like tails. One type doesn't move on its own.

Protozoa are the cause of several serious diseases, including amoebic dysentery, African sleeping sickness, and malaria.

Potent Fact

Worms can grow to lengths of 40 feet and are the largest known disease-causing organisms.

Worms, Worms, Worms

In contrast to protozoa, disease-causing worms are multi-cellular and have complex reproductive cycles that can involve intermediate hosts, such as snails, mosquitoes, and flies, for larval stages before maturing in a human host.

Roundworms

Nematodes, or roundworms, are long, cylindrical worms with unsegmented bodies. A classic example of a nematode is the tiny white parasite called *Wuchereria bancrofti*, which causes elephantiasis, a grossly disfiguring condition that typically involves massive swelling of the legs and other extremities. The worm develops in the mosquito and, as in malaria, is transmitted to humans through mosquito bites. Larvae mature into threadlike adults about $1/2$ inch long. The adults live in the lymph glands and affect drainage, thus causing the disfigurement.

Roundworms can live in many different parts of the body. The roundworm that causes a disease called river blindness lives in the tissue of the eye and the skin. This is a curable disease with simple treatment, but it often occurs in parts of the world where diagnosis and treatment are not common.

Parasites.

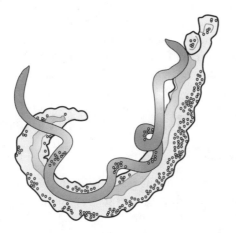

Roundworms also infect other animals, including dogs and cats. It is possible for people to get infections from their pets, so it's important to be sure that all pets are properly wormed and have regular veterinary visits.

Hookworms

Hookworms are another form of roundworm, but they don't have an intermediate host. They infect humans through contact with soil. Hookworms feed on blood, and some can live for up to 15 years.

Hookworm eggs hatch in the soil and infect humans by burrowing through the soles of the feet. The larvae migrate to infect the heart and lungs before passing into the respiratory system and the small intestine. The infection usually starts with skin itch. When the organisms move to the lungs, there may be a cough and a sore throat, but often there are no symptoms. When the worms move to the intestine and feed on blood, anemia may result.

Potent Fact

Wearing shoes and gloves while gardening doesn't only keep feet and hands clean. It can also help to protect against infection.

These infections are curable with one dose of a drug called mebendazole.

Flatworms

Flatworms cause a number of diseases that are transmitted by snails. These worms have a complicated life cycle that involves various species of land and water snails. The most

Antigen Alert

Some flatworms can regenerate if they are cut in half. Tapeworms are flatworms, and this is why they are so hard to eliminate from animals—because if you don't get the whole thing, it can regrow itself and continue to cause infection.

Infectious Knowledge

Stanley Pruisner, Ph.D. received the Nobel Prize in 1997 for his discovery of prions. Some researchers argue that prions are not the cause of disease.

significant of these infections is schistosomiasis, which infects more than 200 million people worldwide. The worm eggs are passed in the urine or feces of the snail, and they hatch in natural waters. Cells get into people through the skin when it comes in contact with contaminated water. The worms migrate to the liver and cause chronic liver, spleen, and bladder damage.

Prions: The Smallest Disease-Causers

Prions are proteins that exist in the brains of all mammals. The normal function of prion proteins is not understood, but recent research suggests that they protect the brain against dementia and other degenerative problems associated with old age. Most researchers agree that prions cause mad cow disease and Creutzfeld-Jakob disease in people, but there are still some who don't accept the theory of prions as infectious agents.

Some prion disease is inherited. However, mutant prions are also capable of turning into rogue disease-causing agents. Transmitted from an infected animal or human to a new host, they convert normal prions they encounter in that host into copies of themselves. Like a rotten apple, once they get inside the brain, the mutant form of prion protein turns the host protein into more copies of the deviant, infectious form. The result is a loss of motor coordination, dementia, a brain full of holes like a sponge, and death, typically following pneumonia. The disease is called mad cow in cattle and Creutzfeld-Jakob (CJD) in people.

Transmission depends on the rogue protein being close enough in structure to the host protein to be able to "lock into" it and convert it. Transmission works best between animals of the same species. For example, a prion disease called kuru found in the Fore tribe of New Guinea is spread by ritual cannibalism, in which mourners eat the brains of their dead relatives. Once this practice was stopped, prion disease rates fell dramatically.

There is evidence that rogue prions can jump from one species to another, provided that the prions of the donor and host are similar enough for the conversion process to occur. For this reason, some people avoid eating meat products of cattle or sheep. For more information on mad cow disease and CJD, see Chapter 18.

Infectious Knowledge
Mad cow disease was identified in Britain in the mid-1980s. Since then, millions of cattle have been slaughtered and dozens of people have died from the related brain-wasting disease called Creutzfeld-Jakob (CJD). Mad cow scares spread across Europe, and governments have had to cope with farmers who lost everything and people who are afraid of this disease that causes a horrible death.

Infectious diseases are caused by a wide variety of organisms from free-living bacteria to viruses that need the machinery of living cells in order to survive. Organisms cause infection in many ways, and the course and severity of an infection depends on its cause as well as our body's response to it.

The Least You Need to Know

◆ Many infectious diseases are caused by organisms that need a host to complete their life cycle.

◆ Viruses are effective disease-causers because they evolve quickly and are difficult to treat.

◆ Other parasites need intermediate hosts, like flies, mosquitoes, or snails, in order to live and cause disease.

◆ Although there is strong evidence that they cause disease, there is still some doubt among scientists about prions.

Chapter 4

The Immune Response: How Our Bodies Fight Infections

In This Chapter

- ◆ The five stages of infection
- ◆ The body's first line of defense
- ◆ The adaptive immune response—disease-specific defense
- ◆ Antibodies vs. antigens
- ◆ The immune system's memory

We are exposed to thousands and thousands of infectious agents every single day. Fortunately, most of us only get sick from them on rare occasions. Surface barriers, like our skin and mucous membranes, are the body's first lines of defense. If the foreign invader gets past the first lines of defense, it might take a while for the body to "figure out" it's been taken hostage by a big bad bug.

Once the body gets wind of the invasion, though, it launches what is called an immune response, which helps to repel the invader. In this chapter, you will learn about the army of cells and chemicals your body uses to respond to an infection and rid the body of the invading organisms.

The Infection Process

The time, course, and severity of an infection depend on the balance between the invader's ability to cause disease, its *virulence*, and the immune system's ability to fight it. As noted in Chapter 3, diseases are caused by bacteria, viruses, fungi, protozoa, and prions. Each has a way that it is transmitted, a way that it divides and spreads, a way it causes disease, and a specific immune response that it triggers in the body.

The five stages of infection are …

◆ **Stage One** The first stage of infection is the incubation period, which is the time between when we are exposed to an infection and when symptoms appear. Although there are no symptoms during this stage, the invader can already be causing substantial damage.

◆ **Stage Two** This stage is when we have nonspecific signs and symptoms such as headache, fever, and tiredness. At this point, it is difficult to know what we have, but we know that we are getting sick. Our bodies have already launched the beginnings of the immune response.

◆ **Stage Three** The acute stage, when specific symptoms show themselves. It's only when infection has reached the acute stage that a doctor can figure out what is wrong based on symptoms. At this point, the immune system is in full swing fighting the invader. This stage determines whether an infection can be naturally resolved, requires intervention (e.g., antibiotics, surgery), or will progress to death.

◆ **Stage Four** A period of resolution in which, if we survive stage three, symptoms gradually go away.

◆ **Stage Five** Finally, there is a period of convalescence when the symptoms are mostly gone, but the body is still recovering.

After these stages, if our immune system and the medications we have taken to help it along have been effective, we are cured.

Disease Diction

Virulence is the disease-causing power of a microorganism. The more virulent an invader, the more likely it is to cause disease.

Potent Fact

The incubation period for diseases is the time between the entrance of the infectious agent and the first appearance of symptoms. It varies dramatically—sometimes it's a matter of hours, sometimes days, and in other cases, like with tuberculosis, it can be years.

Infectious Knowledge

Some organisms live in our bodies peacefully most of the time, but under certain circumstances can cause disease. The fungus *Candida albicans* is an example of this. Given the right conditions, it causes infections that can range from easily curable yeast infections to dangerous infections that spread throughout our bodies and are difficult to treat and cure.

Our First Lines of Defense

The skin and the linings of our digestive, respiratory, and urinary tracts are the body's first line of defense against infection. These physical barriers produce chemicals that inhibit the growth of invaders. For example, the acid that is produced by the lining of our stomachs can neutralize many invaders. Similarly, the mucous lining our respiratory tract prevents many harmful bacteria from entering our system.

In addition, bacteria that live in us all the time and don't cause disease compete for nutrients and attachment sites on cells. Like people rushing for available seats on a crowded subway, these good bacteria take all the seats and leave the invaders with no place to go. *E. coli* is an example of a bacterium that does this.

Disease Diction

Some strains of *E. coli* live harmoniously in our digestive system and even help us to remain healthy. Other kinds, which are often passed on to us in undercooked meats, can cause serious illness.

Infection occurs when an invader crosses these physical barriers. Then our bodies call on the highly specialized forces of the immune system to help repel the invader.

The immune system's response has two parts. First comes the innate response (also called the nonadaptive immune response), which starts immediately and is the same no matter who the invader is. Second is an adaptive response, which takes several days to launch and is specifically designed to attack the organism that is making us sick.

Antigen Alert

Smoking can limit the ability of the mucous membranes in the respiratory tract to provide an effective first line of defense against disease.

Innate Immune Response: The Body's Foot Soldiers

The innate immune response is the same in all people, and it is the same no matter what kind of organism is causing an infection. It does not lead to lasting immunity against any diseases, but it can repel an infection or hold it in check until our body mounts a more specific adaptive response.

Once the body realizes that there is an invading organism in its midst, it first sends special "cell eaters" to the site of the infection. These large white blood cells engulf infectious particles and digest them. They also secrete chemicals called *cytokines*, which signal the body to start the adaptive response.

As cytokines are secreted, they help to increase our body temperature. A high body temperature is good for fighting off infections, because most disease-causing organisms grow

Disease Diction _____

Cytokines are chemicals produced by cells in the immune system in order to communicate and coordinate our body's attack on infectious invaders.

Potent Fact _____

Although fevers make us feel bad, they are our body's way of letting us—and our doctor—know that our immune response has begun. In fact a higher body temperature helps make the immune response better and stronger, so even if we feel bad, having a fever can be a good sign.

This diagram shows how antibodies are created specifically to attach to different antigens in order to inactivate them.

better at lower temperatures, and the adaptive immune response is more intense at higher temperatures.

The cell eaters also secrete a variety of proteins and other chemicals that lead to an inflammatory response. This includes pain, redness, heat, and swelling at the infection site. Although it may be uncomfortable for us, it is the body's way of letting us know that the fight against the infectious invader has begun. During this phase, more cell eaters are recruited to the infection site.

Special Forces in Action: The Adaptive Immune Response

Adaptive immunity is triggered when an infection escapes our innate defenses and makes enough of a substance called an _antigen_ for the immune system to recognize and respond to in a very specific way. Until the infectious agent creates enough antigens for our body to recognize it, we are fighting an unknown enemy.

In response to the antigen, the body makes its own specific chemicals, called _antibodies_, that attach to antigens and inactivate them.

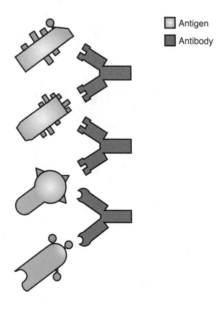

☐ Antigen
■ Antibody

Disease Diction

An **antigen** is a chemical produced by an infectious invader that lives on the surface of the invader and prompts the immune system to produce another chemical that can attach to it and thereby destroy the invader.

An **antibody** is a chemical produced by the body's immune system that can destroy invading organisms. Each time we get a new infection, our body produces specific antibodies necessary to fight it.

There are two kinds of adaptive immunity:

♦ **Cellular immunity** is the part of the adaptive immune response that involves the body's production of specific white blood cells. Some of these cells attack and destroy invaders directly. T lymphocytes or T cells are responsible for cell-mediated immunity. They also orchestrate, regulate, and coordinate the overall immune response. T cells depend on unique cell surface molecules called the major histocompatibility complex (MHC) to help them recognize foreign antigens.

♦ **Humoral immunity** involves the production of antibodies. B cells produce antibodies that circulate in the blood and lymph streams and attach to foreign antigens to mark them for destruction by other immune cells.

Cellular Immunity: Killer White Blood Cells Spring Into Action

The *cellular immune* response is controlled by white blood cells. There are two different types of white blood cells that are able to recognize and destroy microorganisms: B cells and T cells. Both B and T cells originate in the bone marrow. B cells produce antibodies. T cells attack the antigens of invaders directly and control the immune response.

Potent Fact

White blood cells are the soldiers of the bloodstream. They are continually on the lookout for signs of disease. When the body is infected, they travel to the infection site and spring into action. Some—B cells—produce protective antibodies to overpower the invader; others—T cells—surround and devour the invader. White blood cells have a short life cycle, from a few days to a few weeks. A drop of blood has 7,000 to 25,000 white cells, but that number increases if the immune system is fighting an infection.

Infectious Knowledge

It is estimated that 200 species of microorganisms normally live in and on our bodies. Some are permanent, some are just with us for a short while. Over 100 million organisms can live on a square inch of our skin. Our bodies excrete between 100 billion and 100 trillion organisms each day, leaving the total number that live in us so large it has too many zeros to print in this book!

White blood cells fight infection in four ways:

♦ Within several days, the body manufactures T and B cells that are made specifically to respond to the antigens of the invader.

♦ T and B cells multiply and migrate to patrol the tissues of the body. They circulate in blood and in a specialized system of vessels called the lymphatic system. They come in a variety of forms that have different ways of killing invaders.

♦ Cytotoxic T cells come into direct contact with infected cells and kill them.

♦ Helper T cells serve a regulatory function. They are needed to activate B cells and other T cells so they can kill invaders. These helper cells also turn off the T and B cells when their killing job is done.

The organs of the immune system.

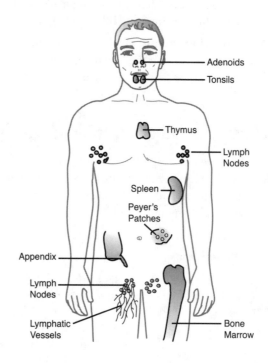

Disease Diction

The **lymph system** is composed of the organs of the immune system that are spread throughout the body. They are responsible for the growth and deployment of lymphocytes—the white blood cells that are the key operators of the immune system. The organs are connected by a network of vessels that are similar to blood vessels. Lymph is a clear liquid that bathes body tissues. Both immune system cells and infectious invaders can travel in the lymph system.

Lymph nodes are small, bean-shaped structures that exist throughout the body. They have compartments where immune cells gather and come into contact with antigens. Lymphocytes are made in the bone marrow from stem cells.

Humoral Immunity: Think Lock and Key

We have talked about antigens, or proteins, that the invader produces. The humoral response is the body's recognition of those specific antigens and its manufacture of antibodies. Think of it as a lock and a key. Each lock—antigen—is slightly different. The keys—antibodies—have two parts: One part is the same in all antibodies, like the rounded part of the key you hold in your hand when opening your front door. The lower part—the part of the key that goes in the lock—varies. Like a key, this part of the antibody can take an almost infinite variety of slightly different forms that allow it to attach to an equally vast number of antigens.

The body recognizes each lock and makes a specific key that fits it in response. Because the invader makes many copies of the lock, the body makes many copies of the key that fits the lock (our body's version of an arms race!). Once the proper key is made, it attaches to the antigen of the invader and stops the invader from multiplying and spreading.

Antibodies are very specific. In HIV, the antigens secreted by the virus change very quickly, so the antigens are able to escape the antibodies our body makes to destroy them. Often, by the time the key (the antibody) is made, HIV has changed the lock (the antigen).

Potent Fact

Collectively, the set of antibodies our bodies produce are called **immunoglobulins**. Immunoglobulins are like memory files of our body's prior exposure to disease.

Immune System Memory: Long-Lasting Protection from Reinfection

Memory is one of the most important biological consequences of adaptive immunity.

Immunological memory is the ability of the immune system to respond rapidly and effectively to invaders it has seen before. The memory is created and stored in the white blood cells and antibodies that were produced during the first infection. These memory cells live a long time and continuously recirculate and patrol the bloodstream. If they come in contact with a previous invader, they jump into action within 24 hours.

Vaccines are created using the principles of immunological memory. When the body is injected with a genetically engineered version of the invader, it launches an immune response and creates memory cells against that invader. Later, the immune system "remembers" previous encounters with infectious organisms and reacts more rapidly and effectively to fight them than it did the first time it "saw" them.

> **Infectious Knowledge**
>
> Although babies are born with weak immune systems, they are protected from some infectious diseases for a few months after birth because they inherit antibodies from their mothers.

Although this chapter makes it clear that our bodies have developed sophisticated ways to fight infection, many infections are hard to cure without outside help. Fortunately, over time, we have discovered a variety of ways to boost our immune system and enhance our own biological efforts to fight infections.

The Least You Need to Know

◆ Most of the time, our bodies are able to fight off invaders without us getting sick.

◆ If invaders take hold, the immune system fights back in a variety of ways.

◆ The first part of the immune response happens quickly, and it is always the same. The second part takes 7 to 10 days and is specifically designed to attack the organism that is making us sick.

◆ Memory cells help us acquire immunity to future infections by identifying organisms that our bodies have seen before.

Chapter 5

Diagnosis and Treatment: Your Doctor's One-Two Punch Against Infection

In This Chapter

♦ How infectious diseases are diagnosed

♦ Drugs that treat and cure infectious diseases

♦ Preventing infectious diseases via vaccines

Usually the body's own immune system is enough to kill biological invaders and make us better. But other times our own immune system just isn't strong enough. Fortunately, scientists and physicians have found a variety of ways to give the immune system a helping hand in treating and curing diseases. The key to any cure or treatment is diagnosis. It's only after a disease is accurately diagnosed that appropriate treatment can begin.

Diagnosis: The First Step to Getting Better

Your throat is sore. You have a cough and a runny nose, maybe a slight fever. It's not awful, but you don't feel so good. What do you do? Call the doctor and ask for medicine? Go see the doctor? Take over-the-counter cold medicine and hope it works? Although over-the-counter medicine may work, if the symptoms persist for more than a couple of days, going to the doctor is probably a good idea to make sure your illness isn't serious.

If you go to the doctor, she may take a swab of your throat and take what's called a culture to see if you have strep throat. Or maybe she will take an x-ray to check for bronchitis or pneumonia. Some tests can be completed immediately; others take a day or more. Some may need to be sent to an outside laboratory for analysis, and the results may take several days. Some tests require blood samples, urine samples, or even stool samples.

The doctor determines which tests to do based on your symptoms. She uses the results of the tests to try to figure out what you have and how to treat it. If the doctor tells you that you have a cold caused by a virus, you will probably be sent home without medicine (remember, viral infections can't be treated with antibiotics). Over-the-counter treatments, rest, and lots of fluids will help until your body fights off the infection. If, however, you have bronchitis or strep, caused by bacteria, you'll get an antibiotic to help you fight off the infection.

> **Potent Fact**
>
> Sometimes a doctor will treat a patient based solely on what he sees, until test results come back. In this case, take your medicine and wait for the test results to come back. Once your doctor knows what the cause of your illness is, he may change your treatment.

There are a variety of ways to diagnose infectious diseases. Many involve looking for the immune system's reaction—the formation of antibodies. As we learned in Chapter 4, it takes a minimum of 7 to 10 days after initial infection for the immune system to launch an antibody response. This means that there is a period after initial infection when an antibody test may be negative even if you have a disease. When this happens, the result is called a *false negative*.

> **Disease Diction**
>
> A **false negative** result occurs when the body hasn't responded to an infection with the production of antibodies, so tests indicate there is no disease present when, in fact, there is.
>
> A **false positive** occurs when a disease has been present in the body before—either from a vaccination or regular exposure that the body has fought off through the production of antibodies. Because the body has the antibodies for the disease, a test for it will come up positive even though no infection is present.

Two major tests are used to detect the presence of antibodies. One is the ELISA (enzyme linked immunoassay test) and the other is called a Western Blot. Both tests use blood samples. The ELISA test is quite rapid, but there is a danger of *false positives* because the test is not necessarily specific—in other words, sometimes it finds antibodies, but not the antibodies that say for sure that you have the particular disease you're being tested for.

The Western Blot test is sometimes done to confirm a positive ELISA because the Western is more sensitive and more specific.

Antigen Alert

Sometimes a recent flu shot can cause a false positive on an ELISA test for HIV.

Why Have Two Tests?

Let's say a patient has been tired a lot and has been battered by a series of colds and infections over the past six months. The patient's doctor, just to be sure, wants to test for HIV, the virus that causes AIDS. (See Chapter 7 for more on HIV/AIDS.) First, the doctor orders an ELISA test, which comes back positive. According to the ELISA, the patient's body is producing antibodies to fight off proteins similar to those found in HIV.

Based on the results of the ELISA alone, can the doctor confirm with 100 percent accuracy that the patient has HIV? No. That's because some proteins found in HIV are similar to proteins found in other infectious agents, and the ELISA can't distinguish the two. The doctor informs the patient of the results, cautions him not to lose hope, and orders another test: the Western Blot.

As with the ELISA, the Western Blot tests for antibodies in the patient's blood, but it is more refined and can pinpoint whether the antibodies are meant specifically for HIV antigens. Although sometimes the test comes back indeterminate (meaning that the information provided doesn't warrant either a positive or a negative conclusion), when the test offers a positive or negative result, the doctor can be fairly confident in those results. In this case, let's say that the Western Blot test came back negative for HIV.

The doctor and patient both breathe tremendous sighs of relief. Of course, the patient will probably want to know why the ELISA gave the false positive result in the first place, causing him tremendous anxiety.

"Was it performed incorrectly?" he asks.

The answer is no, it's just that some proteins found in HIV are similar to proteins found in other infectious agents. So just because the blood sample contained a protein similar to one in HIV, it does not necessarily mean that HIV is present.

Antigen Alert

Interpretation of the Western Blot can be complicated, and sometimes the result of the test will be considered indeterminate. If a patient with an indeterminate test for HIV has engaged in high-risk behavior, such as engaging in unprotected sex or sharing needles, another test should be conducted in six months.

For this reason, the Western Blot is more accurate than the ELISA because it can indicate what, exactly, is in the blood. However, the ELISA test is easier and less expensive and can detect infection earlier. In combination, the two tests make a good team.

Cultures: The Art of Growing Diseases

Another way to diagnose disease is to take a sample from the patient—sputum, blood, urine, or a throat swab—and see if a particular bug grows when a small amount of the sample is placed on a culture plate covered with agar (a gelatin containing nutrients) and kept at a controlled temperature. This is done with tuberculosis using a sputum sample, for example, and with strep using a swab of skin from the throat.

With some infections, the doctor will want to know what drugs might work and what drugs aren't going to be effective in treating it. Cultures can be used to determine how susceptible the infection will be to drug treatment. The bug grown in a sample is divided and put on several culture plates, each containing a different antibiotic in different amounts. If the bug grows on a particular plate, the organism and most likely the infection is resistant to that drug. If no bugs grow, the infection is susceptible, and a potential treatment has been found.

Disease Diction

Sputum? Urine? Stool samples? Puh-leeeeze! **Sputum** is just a fancy way of saying deep secretions from the lungs or throat that we spit up for the doctor. **Urine** is the stuff that makes yellow snow. And **stool samples?** Think: Number two.

An example of a culture plate containing a growing microorganism covered with agar.

(© WHO)

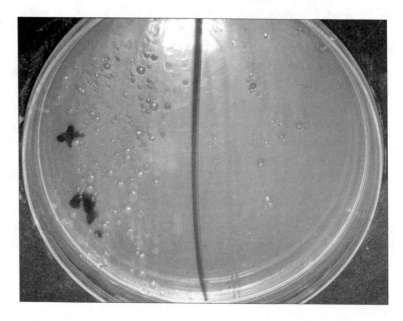

> **Potent Fact**
>
> Why does it take so long to diagnose tuberculosis (TB)? Most bacteria double in 20 minutes, so a culture would become visible in a relatively short period of time. With TB, it takes 24 hours for one bacterium to become two, so it takes a lot longer for the culture to grow enough to be visible on a plate.

The PCR Revolution

You'll recall that the ELISA and Western Blot tests both tested for the presence of antibodies (the good guys) in the patient's blood. It's the presence of antibodies, or the absence of them, which allows doctors to conclude that a patient has or doesn't have a particular infection. But it takes at least 7 to 10 days for the body to produce antibodies that specifically target a particular infection. Why wait, you might be wondering, for the body to produce antibodies? Wouldn't it be better to test for the presence of an infection earlier? After all, the early bird gets the worm.

One way to diagnose disease faster would be to look for something in the infecting organism that identified it, instead of waiting for an antibody response. We know that the genetic material of each living organism has sequences of DNA (or sometimes RNA) that are uniquely and specifically present only in that particular species. These unique variations mean that we can trace genetic material back to its origin and identify the type of organism it came from.

Laboratory worker reviewing a DNA band pattern.

(Courtesy CDC)

PCR is useful for confirming the virulence of anthrax.

(© WHO)

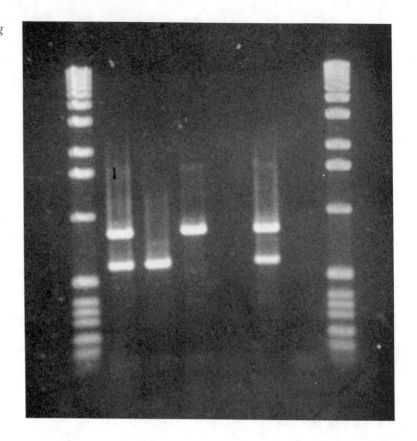

Sounds simple, right? Not exactly. You need a sample with enough copies of that unique segment of DNA that identifies an organism so that it can be detected. Until recently, that wasn't possible. However, about 10 years ago biologist Kary Mullis invented a technique called the polymerase chain reaction, or PCR for short. PCR has revolutionized scientific research and may well revolutionize the way we diagnose disease. It allows scientists and physicians to reproduce enough copies of an organism's DNA to determine what kind of organism it is. (Of course, to diagnose a disease using PCR, you need to know the exact genetic sequence of a unique segment of DNA that identifies the organism that causes the disease.)

Potent Fact

Using PCR, in several hours small samples of DNA, such as those found in a strand of hair at a crime scene, can produce enough copies to carry out forensic tests.

Sometimes called molecular photocopying, PCR can synthesize and analyze any specific piece of DNA or RNA. It can identify and duplicate genetic material from blood, hair, or tissue specimens. It even works on samples that are thousands or even millions of years old!

How does PCR work in diagnosis? It looks directly for unique DNA from a disease-causing organism. If it "finds" the DNA it's looking for, it makes millions of copies of it in less than an hour. Once enough copies are made, scientists can identify the organism. PCR is being used to detect the HIV virus, middle ear infections in children, Lyme disease, and *H. pylori*, the bacterial cause of most stomach ulcers.

Treating Disease ... None Too Kindly

Once your doctor has made a diagnosis, either based solely on symptoms, with a blood test such as the ELISA or Western Blot, with a culture, or using the revolutionary PCR technique, treatment can begin.

Until the twentieth century, there were few known effective treatments for infectious diseases. Once the causes of these diseases were discovered, researchers and physicians began to develop effective ways to fight them. In the past 100 years we have seen the development of antibiotics; the development of a number of successful vaccines and the growth of childhood vaccination programs; and the eradication of diseases like polio and smallpox.

However, it is important to note that we have no drugs that cure viral infections, HIV/AIDS has confounded our ability to develop a vaccine to prevent it, and we aren't even doing that good a job with many of the diseases we know how to treat and cure. Unfortunately, in many parts of the world, the proper treatments we in the United States use as a matter of course aren't readily affordable and available.

Some of the successful treatment methods that have been developed in the past century are discussed in the rest of this chapter.

Antibiotics: Natural Weapons Against Bacterial Diseases

Antibiotics are substances that kill or harm bacteria. Since their discovery in the 1930s, antibiotics have been used to cure many bacterial diseases and have saved millions of lives. When they are used properly, antibiotics are very powerful. Unfortunately, because of their effectiveness, some people use antibiotics too casually, in response to any illness, leading to resistance later.

As noted in the previous chapter, bacteria are living organisms that are constantly changing to increase their chances for survival. Sometimes these changes allow them to resist antibiotics. Antibiotic resistance has become a major public health problem. For this reason, it is important to use antibiotics wisely so that they remain useful to us in our fight against infectious disease.

It's not that scientists and doctors—and the rest of us, for that matter—don't know how to kill bacteria. Many chemicals, including household bleach and cyanide, can knock the lights out of even the most virulent bacteria. The problem with most of the bacteria-killing chemicals is that they kill people, too.

Just as cops wouldn't blow up a whole building full of innocent people to kill the one bad guy hiding out inside, doctors wouldn't use cyanide or bleach—both of which would kill a person—to kill the bacteria hiding out inside of a person's body. Instead, researchers try to find chemicals that will kill bacteria without causing too much harm to us. Fortunately, there are natural products, like penicillin—which comes from a mould—that are effective. Researchers have also found ways to help these natural products perform better, and they have even made some fully man-made, or synthetic, drugs.

Potent Fact

DO … be sure to take antibiotics as prescribed. This includes the time of day, with food or on an empty stomach, and so on. Tell your doctor and pharmacist about other drugs you are taking as well as any known allergies to medication. Take all the antibiotics that you are supposed to take. If you don't, you may not kill all the bugs, and if you get sick again, the drug you took the first time might no longer work.

DON'T … insist on an antibiotic if the doctor says your infection isn't bacterial. Antibiotics only work against bacterial infections. Don't take old antibiotics or leave leftover antibiotics in the medicine cabinet. Throw them away.

Infectious Knowledge

The first antibiotics were made from naturally occurring substances. However, over the past 50 years, researchers also developed antibiotics that are completely man-made. Synthetic drugs are usually just variations of naturally-occurring molecules. Because they can be manufactured by chemical synthesis, they tend to be easier to mass produce in a pure form. Naturally-occurring drugs must be purified from their source in an active form, which isn't always easy.

There are thousands of antibiotics, and they fall into several different classes. Each class attacks a different part of the bacteria. Once a bacterium develops resistance to one antibiotic of a certain class, it is often resistant to all others that work the same way.

How Do Antibiotics Work?

Some antibiotics interfere with the synthesis of the bacterial cell wall, causing a defect in the wall that makes the bacteria burst when it begins to grow and divide. Others keep the bacteria from making specific chemicals needed for cell survival and cell division. Still others cause misreadings of genetic material, which interfere directly with cell division.

The most effective antibiotics are those that actually kill bacterial cells. These are called *bacteriocidal* drugs. Other antibiotics that don't have killing properties are called *bacteriostatic*; these don't kill the bugs but they do prevent the bugs from continuing to grow.

For people with weakened immune systems, bacteriocidal drugs are generally more effective because often their bodies can't help them get rid of an infection.

Sometimes doctors prescribe a combination of drugs to get maximum effectiveness. Using combinations of drugs is also one way to combat the development of resistance.

Some antibiotics work very specifically against disease-causing organisms. These are called narrow spectrum antibiotics. Broad spectrum antibiotics, on the other hand, kill many bacterial cells, including some of the beneficial bacteria that live in the gut. Often the doctor will tell patients to eat a little bit of yogurt that contains live *Acidophilus* cultures every day during a course of antibiotics. This helps protect the "good" bacteria and limits upset stomachs.

> **Disease Diction**
>
> A **bacteriocidal** antibiotic directly kills bacterial cells. A **bacteriostatic** antibiotic prevents the bacteria from multiplying, but doesn't have killing properties.

Treating Viral Infection

We haven't been as successful at treating viral infections as we have bacterial infections. That's because viruses live inside the body's own cells, making them very difficult for the immune system to attack.

Recently, researchers have developed what are called antiviral drugs. One of them, Acyclovir, has been approved in the United States for treating the herpes simplex virus infections, for example. Doctors also use combinations of powerful drugs to fight HIV infections.

Fortunately, many viral infections can be prevented by vaccination. With others, doctors do their best to treat the symptoms.

Dealing with Fungi and Parasites

Antifungal creams and oral and intravenous medications are available for treating both superficial and life-threatening fungal infections, but they aren't always successful and can sometimes have serious side effects. Like the antibiotics that prevent bacteria from growing, antifungal drugs interfere with essential functions of the fungal cell like the cell membrane. However, unlike antibiotics, there are far fewer classes and fewer options for physicians to use for treatment of life-threatening disease. In recent years, potent antifungal medications have been developed for treating skin and nail infections that seem to be more effective.

Unlike many diseases caused by viruses and bacteria, we cannot vaccinate against parasitic diseases. Some parasitic infections can be treated with antiparasitic medications, though they often have serious side effects and are becoming ineffective due to increasing resistance. As with any infectious disease, the best way to handle parasitic infections is to prevent them from happening in the first place.

Prevention *Is* the Best Medicine

Long before we knew how diseases were caused and spread, and hundreds of years before the advent of antibiotics and vaccines, people who recovered from specific diseases didn't get those same diseases if they were exposed again. For example, the Chinese somehow figured out that if they took pus from smallpox lesions of infected people and spread it onto the arms of those who weren't infected, some of those people wouldn't get sick. Although some people did get sick and die, many were protected.

> **Potent Fact**
>
> Pharmaceutical companies are reducing their investment in antibiotic research because there is less money to be made on such drugs than on others, such as lifestyle drugs like Viagra, and because of the lengthy process required to receive FDA approval. Less research will mean fewer new antibiotics to choose from—a dangerous prospect given increasing drug-resistant diseases.

And in 1796, Edward Jenner discovered that milkmaids didn't get smallpox because they were exposed to cowpox, a similar disease that affects cows. This led Jenner to do experiments that showed that exposure to cowpox prevented smallpox. Jenner's process was called vaccination, from *vacca*, the Latin word for cow. The substance used to vaccinate was called a vaccine.

Today, almost all children receive a number of vaccines before they are a year old, and many adults get a flu vaccination every year. The result of vaccination has been a large decrease in diseases that used to kill millions.

How Do They Work?

As we learned in the previous chapter, when we are infected with a disease-causing organism, the body launches an immune response targeted specifically at the invader. A part of this response involves the production of antibodies that are made in response to antigens produced by the invading organism. At the same time, the immune system makes memory cells that stay in the bloodstream. If the body sees the same invader again, the memory cells spring into action quickly to prevent reinfection. Most of the time, memory cells respond fast enough to inactivate the invader so we don't even have any symptoms. In other words, we are immune to reinfection.

A vaccine creates an immune response—the production of antibodies and memory cells—without making us sick. If our body sees the same invader again, it will fight it off automatically.

What's the Recipe?

The first step in making a vaccine is to create a weakened version of an infectious agent, or part of one, that can't cause a full-blown disease but that does contain antigens (the bad guys) that will elicit the creation of antibodies (the good guys) by the body's immune system. There are a number of ways to do this.

Potent Fact

Some diseases that can be prevented by vaccines include the following:

- Anthrax
- Chickenpox
- Hepatitis A and B
- Influenza
- Mumps
- Polio
- Rabies
- Tetanus
- Yellow fever

Killed or inactivated vaccines   With this method, the organism is killed and then injected into the body. The typhoid vaccine and the Salk polio vaccine are examples. Since this isn't the strongest type of vaccine, booster shots may be required every few years to be sure the vaccine continues to work. This type of vaccine can't cause disease, so it is considered safe for those who have compromised immune systems.

Acellular vaccines   These use only the part of the organism that makes antigens. The flu vaccine is an example of this. Like killed vaccines, boosters may be required, and they are safe for those with compromised immune systems.

Attenuated vaccines   This method weakens a live organism by aging it or altering its growth conditions. Vaccines made this way are the most successful, probably because they grow in the body and cause a large immune response. This also means they are the riskiest because they can sometimes cause disease. They are not recommended for patients with weakened immune systems. Examples are measles, mumps, and rubella. Immunity usually lasts a lifetime, so no booster shots are required.

Toxoids   Some vaccines are made from the poisons, also called toxins, disease-causing organisms secrete. The toxins are chemically treated to

Infectious Knowledge

Another method of making a vaccine calls for the use of an organism similar to the one that causes disease. Jenner did this with cowpox, but the most common modern example is the BCG vaccine used to prevent tuberculosis. There is ongoing debate about the effectiveness of BCG, and it is not commonly used in the United States.

decrease their harmful effects. Diphtheria and tetanus vaccines are toxoids. Since the immune response they induce can be weak, they are often given with an adjuvant—another agent that increases immune response. When more than one vaccine is administered together, it is called a conjugated vaccine. Boosters are sometimes required every ten years.

Subunit vaccines Recently, biotechnology and genetic engineering have been used to make subunit vaccines, which use only the parts of an organism that stimulate a strong immune response. Researchers separate the disease-causing genes and then isolate and purify them to be used as a vaccine. The hepatitis B vaccine is an example of this. These vaccines are safe for those with weakened immune systems because they can't cause disease.

> **Potent Fact**
>
> A number of times in the twentieth century the United States government has carried out vaccination campaigns. This happened with polio and smallpox, childhood diseases such as mumps and the measles, and it happens every year with flu vaccines.

The Value of Community Vaccination

Vaccines work well in preventing disease in communities, as well as in individuals. Why? Because infectious diseases require susceptible people in order to spread. Community vaccination cuts down the number of susceptible people in a population, making it harder for diseases to spread. This is the reasoning behind vaccination of kids before they go to school. Community vaccination has helped decrease the incidence of diseases like whooping cough, polio, and smallpox.

So Why Doesn't Someone Make an AIDS Vaccine?

The most effective vaccines use weakened strains of an actual disease-causing organism. Most experts believe that the long-term risks of this approach for an AIDS vaccine would be way too dangerous. In other words, they fear that the weakened strain might not be weakened enough, and would end up infecting people with HIV rather than preventing the spread of the disease.

Researchers have used pieces of proteins from HIV in laboratory tests with great success. Initial optimism changed to great disappointment when experiments showed that vaccinations worked in the laboratory but didn't neutralize viruses found in patients.

Unfortunately, we still don't know why the vaccine doesn't work in patients. One likely reason is that the virus changes very rapidly, even within an individual patient. Once the virus changes enough, the antibodies induced by the vaccine don't recognize it and can't kill it. A second possibility is that a successful immune response requires special killer cells, and such cells aren't generated in response to a vaccine made up of purified or weakened viral proteins.

Hope for the Future

Will there ever be an effective AIDS vaccine? Several approaches are being pursued. One is to make vaccines that contain proteins from many different viral strains. Studies are underway to figure out how many components are needed to get broad protection. Another method is to immunize directly with DNA-based vaccines, which may lead to the production of protective immune cells as well as antiviral antibodies. There are several vaccine trials underway in different parts of the world, but we have yet to find a vaccine that provides broad protection across many populations.

Just as scientists created vaccines to take the pop out of polio and the misery out of mumps and measles, there is hope that they may someday reduce the agony of AIDS with an effective vaccination.

The Least You Need to Know

- Most methods used to diagnose disease involve waiting for the body to produce an antibody response to the infectious invader. This can take days or weeks, so there is a period during early infection when some diseases will be missed.

- New diagnostic technologies allow us to test for genetic material unique to each infectious organism. With these technologies, there is no need to wait for an antibody response.

- Antibiotics are effective against bacterial infections, but not those caused by viruses.

- If antibiotics are prescribed, patients should take them to completion as instructed in order to be sure their infection is completely cured.

- Vaccines are an effective way to prevent disease, but it isn't always easy to make a vaccine that is both safe and effective.

Chapter 6

Epidemics of the Past

In This Chapter

- ◆ Major epidemics in our past
- ◆ Symptoms, death tolls, and early treatments
- ◆ Ancient epidemics that continue to plague us today

Throughout history diseases and epidemics have killed millions of people, destroyed empires, and won wars. In this chapter, you will learn about several of these epidemics and the devastating effects they've had—and can still have—on their victims and those around them.

Smallpox: 12,000 Years of Terror

Smallpox is one of greatest scourges in human history. This disease, which starts with a distinctive rash that progresses to pus-filled blisters and can result in disfiguration, blindness, and death, first appeared in agricultural settlements in northeastern Africa around 10,000 B.C.E. Egyptian merchants spread it from there to India.

The earliest evidence of smallpox skin lesions has been found on the faces of mummies from the eighteenth and twentieth Egyptian dynasties, and in the well-preserved mummy of Pharaoh Ramses V, who died in 1157 B.C.E. The first recorded smallpox epidemic occurred in 1350 B.C.E., during the Egyptian-Hittite War.

In 430 B.C.E., the second year of the Peloponnesian War, smallpox hit Athens and killed more than 30,000 people, reducing the population by 20 percent. Thucydides, an Athenian aristocrat, provided a terrifying account of the epidemic, describing the dead lying unburied, the temples full of corpses, and the violation of funeral rituals. Thucydides himself had the disease, but he survived and went on to write his historic account of the Peloponnesian War. In this work, he noted that those who survived the disease were later immune to it. He wrote, "the sick and the dying were tended by the pitying care of those who had recovered, because they knew the course of the disease and were themselves free from apprehensions. For no one was ever attacked a second time, or not with a fatal result." These Athenians had become immune to the plague.

Athens was the only Greek city hit by the epidemic, but Rome and several Egyptian cities were affected. Smallpox then traveled along the trade routes from Carthage.

In 910, Rhazes (Abu Bakr Muhammad Bin Zakariya Ar-Razi, 864–930 C.E.) provided the first medical description of smallpox, documenting that the illness was transmitted from person to person. His explanation of why survivors of smallpox do not develop the disease a second time is the first theory of acquired immunity.

Potent Fact

Rhazes was a Persian doctor who worked in the main hospital of Baghdad. He ranks with Hippocrates and Galen as one of the founders of clinical medicine and is widely regarded as the greatest physician of Islam and the Medieval Ages. His writings on medicine influenced physicians well through the Renaissance and into the seventeenth century. And his work on smallpox and measles was one of the first scientific treatments of infectious diseases.

The patterns of disease transmission often paralleled peoples' travel and migration routes. Disease in Asia and Africa spread to Europe during the Middle Ages. Smallpox was brought to the Americas with the arrival of Spanish colonists in the fifteenth and sixteenth centuries, and it is widely acknowledged that smallpox infection killed more Aztec and Inca people than the Spanish Conquistadors, helping to destroy those empires.

Smallpox continued to ravage Europe, Asia, and Africa for centuries. In Europe, near the end of the eighteenth century, the disease accounted for nearly 400,000 deaths each year, including five kings. Of those surviving, one-third were blinded. The worldwide death toll was staggering and continued well into the twentieth century, where mortality has been estimated at 300 to 500 million. This number vastly exceeds the combined total of deaths in all world wars.

In the United States, more than 100,000 cases of smallpox were recorded in 1921. Strong declines occurred after that because of the widespread use of preventive vaccines. By 1939, fewer than 50 Americans per year died of smallpox.

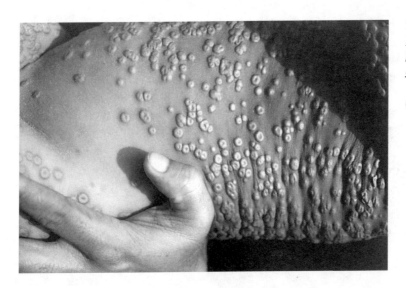

This person, photographed in Bangladesh, has smallpox lesions on skin of his midsection.

(Courtesy CDC/James Hicks)

Variolation: The Earliest Smallpox "Vaccines"

The idea of intentionally inoculating healthy people to protect them against smallpox dates back to China in the sixth century. Chinese physicians ground dried scabs from smallpox victims along with musk and applied the mixture to the noses of healthy people.

In India, healthy people "protected" themselves by sleeping next to smallpox victims or wearing infected peoples' shirts. In Africa and the Near East, matter taken from the smallpox pustules—raised lesions on the skin the contain pus—of mild cases was inoculated through a scratch in an arm or vein. The goal was to cause a mild infection of smallpox and stimulate an immune response that would give the person immunity from the natural infection. This process was called *variolation*.

Unfortunately, the amount of virus used would vary and some would contract smallpox from the inoculation and die. Nonetheless, this preventive approach became popular in China and South East Asia. Knowledge of the treatment spread to India, where European traders first saw it.

An Englishwoman, Lady Mary Wortly Montagu, was responsible for introducing variolation to England. In 1717, while accompanying her husband, the British ambassador to Turkey, in Constantinople she came across the ancient Turk practice of inoculating children with smallpox matter.

Disease Diction

Variolation is the inoculation of matter taken from the smallpox pustules of mild cases through a scratch in an arm or vein. Used by people in the past, the goal was to cause a mild infection of smallpox and stimulate an immune response that would give the person immunity from the natural infection.

Initially horrified at this seemingly savage practice, she learned that a child was protected from the ravages of smallpox through this process. She then had her six-year-old son inoculated while in Turkey, and in 1721, in the presence of Royal Society Members, she had her daughter inoculated. This led to adoption of variolation, mainly by the aristocracy in England and Central Europe. Before long, variolation to prevent smallpox was widespread. During America's War of Independence, George Washington had his army treated in this way. Napoleon did the same with his army before they invaded Egypt.

Edward Jenner: Vaccine Pioneer

During his training as a physician, Edward Jenner learned from nearby milkmaids that after they contracted cowpox they never got smallpox. Cowpox is a far milder disease than smallpox, yet the diseases are quite similar. In 1796, Jenner decided to test the theory that infectious material from a person with a milder similar disease could protect against a more severe disease.

He put some pus from a cowpox pustule on small cuts made on the arm of James Phipps, an eight-year-old boy. Eight days later, Phipps developed cowpox blisters on the scratches. Eight weeks later, Jenner exposed the child to smallpox. The boy had no reaction at all, not even a mild case of smallpox. The cowpox had made him immune to smallpox. Jenner developed the first vaccine, using cow serum containing the cowpox virus. Jenner tried this new treatment on eight more children, including his own son, with the same positive result.

Potent Fact

The word vaccination is derived from the Latin word for cow, *vacca*.

Potent Fact

Amazingly, eradication of smallpox, one of the world's most deadly scourges, cost approximately $100 million. Even in today's dollars, this was a bargain.

After a period of slow acceptance, Jenner's vaccine approach was widely adopted. Vaccination using Jenner's method was key in decreasing the number of smallpox deaths, and it paved the way for global eradication of the disease.

The World Takes Action

In 1959, The World Health Assembly decided to organize mass immunization campaigns against smallpox. The World Health Organization (WHO) announced the global smallpox eradication program in 1967. At that time there were still an estimated 10 to 15 million cases of smallpox a year resulting in two million deaths, millions disfigured, and another 100,000 blinded. Ten years later, after dispersal of 465 million doses of vaccine in 27 countries, the last reported naturally occurring case appeared in Somalia. On October 22, 1977, a 23-year-old male, Ali Maow Maalin, developed smallpox and survived.

The global campaign against smallpox ended in 1979 just two years after Maalin's case. Two additional cases of smallpox occurred in Birmingham, England, in 1978, after the virus escaped from a laboratory. There has not been a case reported in more than 25 years.

Variola: The Cause of Smallpox

Smallpox is caused by a virus and can result in one of two forms of the disease, called *variola* major and variola minor. Variola major kills 20 to 40 percent of unvaccinated people who get it and can lead to blindness. Variola minor, a far less lethal form of the disease, results in death only on rare occasions.

> **Disease Diction**
>
> A sixth-century Swiss bishop named the cause of smallpox **variola,** from the Latin *varius,* meaning "pimple" or "spot." In the tenth century, the term *poc* or *pocca* was used to describe the scars left behind, which resembled "pouches." When syphilis became epidemic in the fifteenth century, the term smallpox was adapted to distinguish between the diseases.

The disease is transmitted primarily by direct contact with droplets from saliva and other body fluids that travel through the air, such as through a sneeze. It may also be spread if an uninfected person handles clothing worn by someone with the disease.

Signs and Symptoms of Smallpox

The incubation period for smallpox is 8 to 17 days, with people usually getting sick 10 to 12 days after infection. Symptoms start with malaise, fever, rigors, vomiting, headache, and backache. The trademark smallpox rash appears after two to four days, first on the face and arms and later on the legs, quickly progressing to red spots, called papules and eventually to large blisters, called pustular vesicles, which are more abundant on the arms and face. Although full-blown smallpox is unique and easy to identify, earlier stages of the rash could be mistaken for chickenpox. When fatal, death occurs within the first or second week of the illness.

There is no effective treatment for smallpox. There are antiviral drugs that might work, but they have not been tested due to restrictions on smallpox research.

Smallpox Vaccine

The smallpox vaccine currently licensed in the United States is made with a virus called vaccinia, which is related to smallpox. It does not contain the actual smallpox (variola) virus. Vaccinia causes the body to produce antibodies that protect against smallpox and several other related viruses.

When a person is vaccinated, the usual response is the development of a red spot at the vaccination site two to five days after the shot. The red spot becomes pustular, and reaches its maximum size in 8 to 10 days. The pustule dries and forms a scab, which separates 14 to 21 days after vaccination, leaving a scar. Sometimes there is also swelling and tenderness of lymph nodes. A fever is common after vaccine. Fatal complications are rare, with less than one death per million vaccinations.

The CDC is the only source of smallpox vaccine and will provide it to protect laboratory and other health-care personnel at risk for exposure. A reformulated vaccine is now under development.

Smallpox: An Agent of Bioterrorism?

Several years ago, Ken Alibek, a former deputy director of the Soviet Union's civilian bioweapons program, indicated that the former Soviet government had developed a program to produce smallpox virus in large quantities and adapt it for use in bombs and intercontinental ballistic missiles.

> **Potent Fact**
>
> There were approximately 15 million doses of 20-year-old vaccine available following the September 11, 2001, terror attacks. However, once bioterrorism in the form of anthrax became a real threat, the United States government urgently ordered another 150 million doses of smallpox vaccine to be made available within short order as a precaution.

If a smallpox vaccine exists, smallpox bioterrorism shouldn't be a problem, right? Wrong. The vaccine program in the United States was so successful that routine vaccination was discontinued in 1972. Nearly 50 percent of the population has never been vaccinated and, of the vaccinated individuals, the vaccine is of questionable value since it requires boosting every 10 years. For the first time in nearly a century, the United States population is at significant risk for smallpox. (See Chapter 20 for more on smallpox and other agents of bioterrorism.)

By international agreement, the main stores of smallpox virus from the Cold War superpowers are kept securely at the CDC's headquarters in Atlanta and at a similar institute in Moscow.

The Black Death: Bubonic Plague

Ring around the rosy,
A pocket full of posies,
Ashes ... ashes,
We all fall down.

A familiar nursery rhyme that children have recited as a harmless play song for generations ironically refers to one of Europe's most devastating diseases. The bubonic plague, better known as the "The Black Death," has existed for thousands of years. The first recorded case of the plague was in China in 224 B.C.E. But the most significant outbreak was in Europe in the mid-fourteenth century. Over a five-year period from 1347 to 1352, 25 million people died. One-third to one-half of the European population was wiped out!

The first symptoms of bubonic plague appeared within days after infection: fever, headache, and a general feeling of weakness, followed by aches in the upper leg and groin, a white tongue, rapid pulse, slurred speech, confusion, and fatigue. By the third day, a painful swelling of the lymph glands in the neck, armpits, and groin occurred, and these enlarged areas were called "buboes." Bleeding under the skin followed, causing purplish blotches. Dark-ringed red spots on the skin from infected fleabites, or "ring around the rosy," eventually turned black, producing putrid-smelling lesions. The victim's nervous system collapsed, causing extreme pain and bizarre neurological disorders. This was the inspiration for "Dance of Death" rituals. Fragile nasal capillaries led to excessive sneezing. By the fourth day, wild anxiety and terror overtook the sufferer. Finally, a sense of resignation registered as the skin blackened, giving rise to "The Black Death." The simplistic words in the famous nursery rhyme capture the essence of plague's horror.

> **Infectious Knowledge**
>
> The excessive sneezing of plague suffers led Pope Gregory VII to coin "God Bless You" as a holy response when someone sneezes.

The rhyme also describes highly aromatic flowers and herbs, the "pocket full of posies," that people carried with them and held near their faces to ward off the horrid odor associated with the plague. Many corpses were uncharacteristically cremated—the "ashes, ashes,"—and finally, death would come, and we would "all fall down."

Plague infects both people and rodents, with rodents helping to transmit it further within the population. Fleas feeding on infected rodents can transmit the disease to people as well. Once infected, people can infect others by coughing, sneezing, or close talking.

The origin of "The Black Death" dates to an outbreak in China during the 1330s. During this period, China was an important trading nation, and international trade via the Silk Road helped create the world's first pandemic. Plague-infected rats on merchant ships spread the disease to western Asia and Europe. In the fall of 1347, Italian merchant ships with crewmembers dying of plague docked in Sicily, and within days the disease spread to the city and the surrounding countryside. The disease killed people so quickly that the Italian novelist Giovanni Boccaccio, whose father and stepmother died of plague, wrote that "its victims ate lunch with their friends and dinner with their ancestors in paradise." By August, the plague had spread as far north as England.

For five years, the disease would disappear in winter, when fleas were dormant, and resume its killing spree each spring. The impact of the plague was enormous, as governments, trade, and commerce virtually ceased. Faith in religion decreased because many clergy died and prayer failed to prevent sickness and death. Because trade was difficult, the prices of goods were grossly inflated. A decimated work force required higher wages, leading to peasant revolts in England, France, Belgium, and Italy.

The disease receded in 1353, but never really went away. Smaller outbreaks continued for centuries, affecting all of European society, rich and poor. Two hundred years of recurring death immeasurably changed government, the arts, science, and religion.

Plague epidemics ravaged London in 1563, 1593, 1603, 1625, 1636, and 1665, reducing its population by 10 to 30 percent during those years.

Infectious Knowledge

The English writer Daniel Defoe, the author of *Robinson Crusoe*, wrote graphically about the plague years 1664–1665. "It is scarce credible what dreadful cases happened in particular families every day. People in the rage of the distemper, or in the torment of their swellings, which was indeed intolerable, running out of their own government, raving and distracted, and oftentimes laying violent hands upon themselves, throwing themselves out at their windows, shooting themselves, mothers murdering their own children in their lunacy, some dying of mere grief as a passion, some of mere fright and surprise without any infection at all, others frightened into idiotism and foolish distractions, some into despair and lunacy, others into melancholy madness."

The Italian plague of 1630 claimed between 35 and 69 percent of the population. The German plagues between 1709 and 1713 were equally devastating, and in 1720, plague reduced the population of Marseille by 40 percent. The bubonic plague is believed to have killed 137 million people during its 400-year reign of terror.

Solving the Plague Mystery

In 1894, two scientists, Alexandre Emile Jean Yersin and Shibasaburo Kitasato, separately described organisms they found in the swollen lymph nodes, blood, lungs, liver, and spleen of dead plague patients. To confirm their findings, they used cultures (see Chapter 5 for more on cultures) taken from patient specimens to infect a variety of animals. All the animals died within days. The same rod-shaped organisms found in the original specimens were also found in the animal organs.

On the island of Formosa, residents considered handling dead rats a risk for developing plague. These observations led P. L. Simond in the late 1890s to suspect that fleas might play a role in the transmission of plague. He observed that people contracted plague only if they were in contact with recently dead rats. They were not affected if they touched rats that were dead for more than 24 hours. Simond showed that the rat-flea transmitted the disease. In a now classic experiment, a healthy rat separated from direct contact with a recently killed plague-infected rat, died of plague after the infected fleas jumped from the first rat to the second.

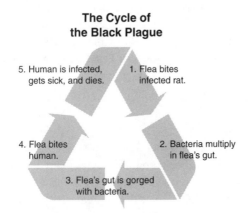

The Cycle of the Black Plague

5. Human is infected, gets sick, and dies.

1. Flea bites infected rat.

4. Flea bites human.

2. Bacteria multiply in flea's gut.

3. Flea's gut is gorged with bacteria.

Life cycle of the Black Plague, as the bubonic plague is sometimes called.

Buboes: Signs of the Plague

Unlike smallpox, the plague is still a threat in some parts of the world. *Yersinia pestis,* the bacterium that causes bubonic plague, is transmitted through rat-tainted fleabites in densely populated cities and in countries with poor hygiene, or in the open country from infected wild rodents. The most common form of human plague is a swollen and painful lymph gland that forms buboes.

Bubonic plague should be suspected when a person develops a swollen gland, fever, chills, headache, and extreme exhaustion, and lives in a region with infected rodents, rabbits, or fleas. Infection of the lungs causes the pneumonic form of plague, resulting in a severe respiratory illness. This form of the disease can spread rapidly and is more highly fatal.

A person usually becomes ill with bubonic plague two to six days after being infected. When left untreated, the plague-causing bacteria invade the bloodstream and kill 50 to 90 percent of people who get it. Antibiotic therapy effectively treats bubonic plague.

Pneumonic plague is more difficult to treat, and even with antibiotics, victims can die from it. Pneumonic plague occurs when the infectious bacteria infects the lungs. The first signs of illness in pneumonic plague are fever, headache, weakness, and a cough that produces blood or watery sputum. The pneumonia progresses over two to four days and, without early treatment, death ensues.

Can I Catch the Plague?

Bubonic plague is still prevalent in more than 20 countries. In the United States, the last rat-borne epidemic occurred in Los Angeles in 1924–1925. Since then sporadic cases have occurred, mostly in western states. Sources of cases today are wild rodents, especially squirrels, prairie dogs, and other burrowing rodents.

Plague is found in parts of Russia and China and regularly occurs in Madagascar. Severe outbreaks have occurred in recent years in Kenya, Tanzania, Zaire, Mozambique, and Botswana. Plague also has been reported in western and northern Africa. In South America, plague is found in parts of Bolivia, Peru, Ecuador, and Brazil. There is no plague in Australia, and Europe has not seen a case for more than 50 years.

Why Is Yersinia So Successful?

A team of scientists recently mapped the entire genetic structure (or *genome sequence*) of *Yersinia pestis*, the plague-causing bacterium. The genome displays many irregularities due to genetic exchange with other microorganisms, and many of its genes appear to have been acquired from other bacteria and viruses. The evidence suggests that plague has undergone large-scale genetic change leading to rapid evolution, which makes it able to adapt to survive in many different environments.

Potent Fact _____

The **genome sequence** is the entire complement of DNA in the cells of an organism. Mapping the genome sequence of the bug that causes plague was valuable because it helped researchers to learn about its evolution. The large number of genetic changes in the organism over time explain its ability to succeed as a disease-causer over hundreds of years.

Genome sequences are becoming available for more and more organisms. The human genome has been sequenced as well. There are many useful pieces of information that will help us fight all kinds of infectious and genetic diseases that come from sequencing of genomes, many of which will be discussed in Chapter 23.

Plague Vaccine

Plague vaccines have been used since the late nineteenth century, but their effectiveness is uncertain. Vaccination reduces the incidence and severity of disease resulting from the bite of infected fleas, but it isn't 100 percent effective. The plague vaccine is licensed for use in the United States and is available for adults at high risk—people who live in the western United States, people who will be in parts of the world where plague is still endemic, and people who are around rodents. Severe inflammatory reactions are common, and plague vaccine should not be given to anyone with a known hypersensitivity to beef protein, soya, casein, or phenol. Finally, the vaccination routine is complex and requires frequent boosters to maintain its effectiveness.

Influenza: A Twentieth-Century Epidemic

On June 28, 1914, Archduke Franz Ferdinand, heir to the throne of the Austro-Hungarian Empire, was assassinated in Sarajevo, starting World War I. Four years and three months later, on November 11, 1918, an armistice was signed in Northern France ending "The Great War." The death toll was enormous, estimated at 8 to 10 million, but it paled in comparison to the *influenza* pandemic that also struck in 1918. War, for all its horror and casualties, was actually less deadly than the outbreak of influenza.

In the spring of 1918, soldiers in the trenches in France complained of sore throats, headaches, and general malaise. Most of them recovered quickly, and only a few died. The soldiers called their illness the Spanish Flu although its origins were, and still are, unknown. By summer of that same year, soldiers' symptoms became much worse. One in five who got sick developed pneumonia or blood poisoning. Many died. Others developed a strange condition called heliotrope cyanosis—they literally turned blue! Almost all of them died within a few days. This second wave of the epidemic spread quickly. More than 70,000 American troops on the Western Front were hospitalized, and one-third of them died.

Disease Diction

Influenza is a common, contagious respiratory infection caused by a virus with outbreaks of different forms occurring almost every winter with varying severity. It is characterized by fever, muscle aches, headache, and sore throat.

The virus did not play favorites in the war, and by the end of the summer, the infection had reached Germany, and over 400,000 civilians died there. The first cases in Britain showed up in Glasgow during May 1918, and in a few months' time the virus killed 228,000 Brits. The epidemic swept through the United States in September, and by early December about 20 million people were infected and 450,000 Americans had died. That was not the worst of it. India suffered the largest toll. The first cases appeared in Bombay in June 1918, followed by cases in Karachi and Madras. Many of India's doctors were serving with the British army, so the country was unprepared to deal with the enormity of their problem. In one year, 16 million people in India died.

The 1918–1919 worldwide epidemic, or pandemic, is estimated to have infected 500 million people resulting in nearly 40 million deaths.

The Flu Today

Today, the influenza virus appears virtually every year in a slightly different form and causes respiratory infections. Some years, cases are sporadic and local. In other years there are epidemics that can spread through cities, rural areas, and even whole countries.

Potent Fact

Acute infections usually have a rapid onset and require medical attention to cure. **Chronic infections** may start slowly and last for a long time. Often medical help manages a chronic infection but cannot cure it.

Each winter, the flu is the most frequent cause of *acute* respiratory illness requiring medical care. It is highly contagious, affects all age groups, and has caused epidemics and pandemics of human disease for many centuries. During most flu seasons, up to 20 percent of the population becomes infected, and approximately 1 percent of those infected have to be hospitalized. Close to 20,000 Americans die from complications of the flu each year. In the United States, the flu pandemics of 1957 and 1968 were associated with an attack rate of up to 50 percent and an estimated 100,000 deaths.

Signs and Symptoms of the Flu

The flu virus spreads through the air. It moves from person to person primarily when someone who is infected coughs or sneezes. The incubation period is one to four days, and you can be infectious, or pass the virus to someone else, starting the day before symptoms begin, and lasting through five days after you get sick. The infectious period for children can be longer. Uncomplicated flu in adults and children is characterized by the abrupt onset of the following symptoms:

- Fever and chills
- Headache
- Sore throat
- Muscle and joint pain
- Severe malaise
- Dry cough
- Runny nose

Antigen Alert

A major new strain of influenza appears approximately every 10 to 15 years, causing a worldwide flu epidemic.

Fever usually peaks within the first day and lasts up to five days.

The flu usually goes away after several days, although some symptoms, like coughing, can last for more than two weeks. In some people, the flu aggravates underlying medical conditions. The risks for complications, hospitalizations, and deaths from flu are higher among people over age 65, very young children, and people with underlying health conditions. During flu epidemics, deaths can result from pneumonia as well as from heart conditions and other chronic diseases.

A Quick-Change Artist

All viruses have genetic material, either DNA or RNA. Viruses replicate—make copies of themselves by using the machinery of their hosts to copy their DNA or RNA—in order to cause infection. During replication, errors in the sequence of the building blocks that make up DNA and RNA can occur, causing the resulting copies of the virus to be slightly different than their "parents." If this happens a number of times, the virus changes enough to elude the antibodies that worked to kill it before.

As the flu virus grows and spreads, the replication of individual cells occurs many times, and the errors compound. Eventually, though the virus remains a flu virus, the cumulative effect of many genetic errors results in a new strain. This is similar to taking a word and rearranging the letters one by one. At first it might be nonsense, but eventually you could have a new word.

If your body becomes infected with this new strain, it's as though it is meeting the virus for the very first time. The fact that you may have had the flu last year makes absolutely no difference. The antibodies your immune system produced in the past give you little or no protection against another flu virus type or subtype. This is why it is not possible to give a vaccine that gives us long-term immunity to the flu. It is also the reason for seasonal epidemics, and for the incorporation of one or more new strains into each year's flu vaccine.

Potent Fact

The United States will produce about 88 million doses of vaccine in 2002. Not all will be taken. In persons over age 65, the vaccination rate is about 65 percent. The vaccination rate for those age 18 to 64 is 31 percent.

Controlling the Flu

Each year the flu vaccine in the United States contains three strains representing the flu viruses most likely to circulate in the United States in the upcoming winter. The effectiveness of the vaccine depends on the age and health of the vaccine recipient and the degree of similarity between the viruses in the vaccine and those in circulation. When the vaccine and circulating viruses are similar, the vaccine prevents illness in approximately 70 to 90 percent of healthy people under 65. Among the elderly who live in nursing homes, the flu vaccine is most effective in preventing severe illness, secondary complications, and death.

Vaccinating people at high risk for complications before the flu season each year is the most effective means of reducing the flu's impact. When vaccine and epidemic strains are well matched, vaccination of health-care workers and others in close contact with people in groups at high risk—or among people living in closed settings like nursing homes or other chronic-care facilities—can reduce transmission and subsequent complications.

In view of its enormous human and economic toll, the flu remains a major target for improved vaccines, vaccine delivery, and antiviral treatment. The use of flu-specific antiviral drugs for treatment is an important adjunct to vaccine. However, these agents are not a substitute for vaccination.

> **Infectious Knowledge**
>
> In Russia there is a flu vaccine that is given through the nose instead of by injection; this vaccine is under development in the United States. It is possible that the use of a vaccine without an injection may encourage more people to get flu shots each fall.

Smallpox, plague, and influenza are just three examples of epidemics that have struck throughout history. They and other infectious diseases have caused devastation and death all around the globe. Although smallpox has been eradicated, there is still fear about its use as a bioterrorist weapon, and most other infectious diseases are still causing illness despite the implementation of better diagnosis, treatment, and prevention.

The Least You Need to Know

- Infectious disease epidemics have caused widespread devastation and death throughout recorded history.
- Most infectious agents continue to cause disease after thousands of years.
- Smallpox is the only disease that has been eradicated, and there are still stores of it kept in the United States and Russia.
- Although vaccines are helpful in fighting diseases caused by viruses, they are not always effective.

Part 2

Raging Epidemics

If the CDC were to issue a Most Wanted list, pictures of the organisms responsible for causing the diseases discussed in this section would be plastered on the walls of doctors' offices across the country.

HIV, the virus that causes AIDS, one of the scariest and most deadly diseases of the past 20 years, would probably top the list. Close on its heels you'd find the organisms causing tuberculosis, a disease we thought we'd given a life sentence once before; malaria, who with its cronies—mosquitoes—hides out mostly in tropical regions; the nasty gang of sexually transmitted diseases that take advantage of people when they least expect it; and hepatitis, a sneaky bug that is silently harming tens of millions of people around the world.

Chapter 7

A Modern Pandemic: HIV/AIDS

In This Chapter

- ◆ History and discovery of AIDS
- ◆ Symptoms, diagnosis, and treatment of AIDS
- ◆ Impact of AIDS worldwide
- ◆ The search for an effective vaccine

The June 5, 1981, Centers for Disease Control and Prevention's (CDC) weekly report of disease trends noted an unusual pattern of opportunistic infections in five gay men. The report began …

> In the period October 1980–May 1981, 5 young men, all active homosexuals, were treated for biopsy-confirmed *Pneumocystis carinii* pneumonia [a fungal infection of the lungs] at 3 different hospitals in Los Angeles, California. Two of the patients died. All 5 patients had laboratory-confirmed previous or current cytomegalovirus (CMV) infection and candida mucosal [yeast] infection.

The report ended with a poignant editorial note:

> *Pneumocystis* pneumonia in the United States is almost exclusively limited to severely immunosuppressed patients. The occurrence of pneumocystosis in these 5 previously healthy individuals without a clinically apparent underlying immunodeficiency is unusual. The fact that these patients were all homosexuals suggests an association between some aspect of a homosexual lifestyle or disease acquired through sexual contact and *Pneumocystis* pneumonia in the population ....

> All the above observations suggest the possibility of a cellular-immune dysfunction related to a common exposure that predisposes individuals to opportunistic infections such as pneumocystosis and candidiasis ...

The age of AIDS had begun!

The Birth of a Disease

At the same time that *pneumocystosis* was occurring in gay men in Los Angeles, *Kaposi's sarcoma*, a rare cancer, was found in a larger number of gay men in New York. Heterosexual drug users, both men and women, were also getting the same two infections. While some blamed illicit drug use, others focused on an infectious disease link. The name "Gay Cancer" was soon replaced by "Gay-Related Immune Deficiency" or GRID. Haitian refugees in Miami and hemophiliacs were also linked to this new syndrome.

Disease Diction

Kaposi's sarcoma is a type of cancer that men with AIDS may develop. It is rarely seen in women. Kaposi's mainly affects the skin, mouth, and lymph nodes, but it can also involve the bowels and lungs. Kaposi's growths, called lesions or tumors, can show up in a wide range of colors. The lesions can appear anywhere on the body and may look like other skin lesions, so a biopsy is generally needed to be sure of the diagnosis. The skin lesions are usually flat and painless and they don't itch or burn. If the disease spreads throughout the body, chemotherapy may be necessary to treat it.

Pneumocystis pneumonia (PCP), is caused by a microscopic fungus that lives in the lungs. Most PCP cases are in people with weakened immune systems, like AIDS patients, cancer patients, and transplant patients. Symptoms include fever, cough, and abnormal breathing. This is the most common pediatric illness associated with AIDS.

It soon became clear that an infectious agent, probably a virus, was spreading through blood. By the end of 1981, 422 cases were diagnosed in the United States and 159 people were dead. In 1982, the numbers rose to 1,614 cases diagnosed and 619 dead. The same year, the acronym GRID was replaced with AIDS for "acquired immune deficiency syndrome" because the disease, which is acquired from someone else, results from an inability or deficiency of the immune system to work properly.

By early 1983, evidence was building that the agent causing AIDS could be transmitted sexually as well as through blood and blood products. Certain groups were at increased risk for disease. At the end of 1983, 4,749 cases were reported and 2,122 were dead.

As the epidemic grew, there was serious concern about the safety of the blood supply. People who received transfusions and hemophiliacs receiving a specific clotting factor were getting AIDS. Unfortunately, health officials miscalculated the magnitude of the problem and during the early 1980s almost half the 16,000 hemophiliacs in the United States contracted AIDS. It was not until 1985, after HIV (human immunodeficiency virus), the cause of AIDS, was isolated and characterized that an antibody-based test to determine whether blood was safe was developed and approved. Much has been written about this sad early chapter of the AIDS epidemic.

Potent Fact

The CDC's National AIDS Hotline Number is 1-800-243-7887.

From Epidemic to Pandemic

As the AIDS epidemic grew unabated in the United States, numerous cases appeared in Europe, South America, and Africa. While disease transmission was generally associated with gay men in the United States, disease transmission in other parts of the world was linked to heterosexual contact. In 1990, the World Health Organization (WHO) estimated that about one million people were living with AIDS. In less than 10 years, HIV had exploded worldwide.

Infectious Knowledge

AIDS, acquired immune deficiency syndrome, is an infectious disease caused by the human immunodeficiency virus. AIDS is characterized by the appearance of opportunistic infections such as tuberculosis, fungal infections, meningitis, and syphilis. Weight loss and wasting are also effects of AIDS. The virus attacks immune system cells, and T cell counts and viral load counts are also important in both diagnosis and determination of appropriate treatment.

There is no cure for AIDS, but there are drugs that can treat the symptoms and infections that go along with the disease.

In the early and mid-1990s, extensive education programs in North America and Europe, stressing the need for protection from bodily fluid transfer, helped to slow the spread of AIDS. Yet, around the world, the number of AIDS cases continued to rise. According to WHO, in 2001 an estimated 40 million people, 37.2 million adults and 2.7 million children, were living with HIV. Twenty years after the first cases were reported, a modern pandemic is ravaging the world's poor.

Most striking are the 28.1 million AIDS cases in Africa, with most in sub-Saharan Africa. Nearly 2.3 million Africans died from AIDS in 2001, and without adequate treatment and care, many more will die. Other countries, like India, are experiencing meteoric rises in AIDS cases. Some countries with little history of AIDS, like China and Russia, are poised for a major epidemic as well.

Clinical Definition of AIDS

The, CDC defines AIDS in an adult or adolescent aged 13 years or older as the presence of one of 25 conditions that indicate that the immune system is severely compromised because of HIV infection. People with HIV are also defined as having AIDS once their CD4 + T cell count goes below 200 cells per cubic milliliter of blood (normal CD4 counts are between 500 and 1600 per cubic milliliter).

AIDS refers only to the end stage of a progressive disease that starts with HIV infection and leads to severe impairment of the immune system. Once the immune system is weakened, AIDS patients are susceptible to other infections and cancers. These infections are called opportunistic because the organisms that cause them take advantage of the person's weakened immune system and cause disease.

> **CAUTION**
> **Antigen Alert**
>
> T cell count is considered a critical measure of health because it tells whether the immune system is working properly. When our T cell count goes down, so does our ability to fight infections.

> **Disease Diction**
>
> A retrovirus is a type of virus with RNA as its genetic material. These viruses can convert their RNA to DNA and insert themselves into their host genome. Once inserted, they can stay latent for long periods of time and reactivate by transcribing copies of themselves at any time.

HIV, the Cause of AIDS

In 1983, Dr. Luc Montagne and colleagues at the Institute Pasteur in France reported that they had isolated a new virus called lymphadenopathy-associated virus, or LAV, which was the cause of AIDS. A short time later, in 1984, Dr. Robert Gallo and colleagues at the National Cancer Institute (NCI) reported that they had isolated a virus called HTLV-III, which was believed to cause AIDS. Both were right, as the virus from French and American groups were nearly identical. Sometime later the virus was renamed human immunodeficiency virus, or HIV. It belongs to the family, or group, of *retroviruses* called lentiviruses, which are found in a wide range of nonhuman primates.

A retrovirus's genes are made up of RNA, and so they need to live in a host and "borrow" the host's DNA in order to multiply.

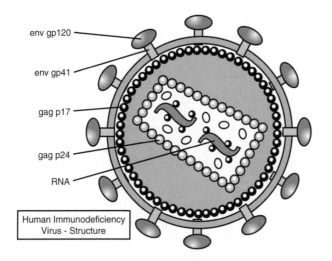

env gp120

env gp41

gag p17

gag p24

RNA

Human Immunodeficiency
Virus - Structure

The mature HIV is roughly spherical. The outer envelope is studded with two major proteins (gp120 and gp41). The central core contains viral proteins (p24 and p17), two copies of the HIV RNA genome, and three viral enzymes essential for viral replication.

The Koch Test

Nearly all scientists recognize that HIV is the cause of AIDS. However, a small and vocal minority of researchers have argued that AIDS is not consistent with a single disease caused by a single infectious agent.

As mentioned in Chapter 2, German bacteriologist Robert Koch (1843–1910) developed a series of rules, or postulates, for deciding whether a microbe is responsible for a specific disease. These four postulates stipulate that an infectious agent must be found in all cases of the disease; the agent must be isolated from the host's body; the agent must cause disease when reintroduced into a healthy host; and the same agent must once again be isolated from the newly diseased host. As you'll see in the following examples, all four postulates have been fulfilled for HIV, convincingly identifying HIV as the cause of AIDS.

♦ In the mid-1980s, laboratory workers studying HIV in isolation were accidentally exposed to high concentrations of the virus and later developed weakened immune systems and opportunistic infections characteristic of AIDS. HIV was isolated from the lab workers' blood, and it was shown to be the same infecting strain. The same observations have been made for health-care workers accidentally infected by puncture wounds from HIV-contaminated needles.

♦ The development of AIDS following known HIV exposure has been routinely observed in blood transfusion cases, mother-to-child transmission, injection-drug use, and sexual transmission. In these cases, the identification of HIV in the blood has been documented using blood samples taken before and after infection.

♦ HIV-infected mothers have been known to give birth to twins where one twin is HIV-positive and develops AIDS, while the other child shows no sign of HIV and remains clinically and immunologically normal.

♦ Numerous studies of HIV-infected people show that the amount of HIV in the body correlates with the progression of the clinical symptoms of AIDS.

♦ Nearly everyone with AIDS has antibodies to HIV.

♦ Before the appearance of HIV, AIDS-related conditions such as *Pneumocystis carinii* pneumonia (PCP), Kaposi's sarcoma (KS) and infection with *Mycobacterium avium* complex (MAC) were extraordinarily rare in the United States.

♦ HIV has been repeatedly isolated from the blood, semen, and vaginal secretions of patients with AIDS, which is consistent with findings linking AIDS transmission via sexual activity and contact with infected blood.

CAUTION

Antigen Alert

It takes several weeks for the body to respond to HIV infection and produce antibodies to fight the infection. Therefore, an antibody test to diagnose HIV will show a false negative if the infection is brand new.

Infection and Disease

Infection begins when an HIV particle encounters a specialized immune cell, a T cell. The virus particle attaches itself to the cell's outer surface membrane and enters the cell. Once in the cell, the virus releases its RNA, and a chemical converts it into HIV DNA. The new HIV DNA moves into the cell's nucleus and strong-arms the host cell to help make copies of the virus, releasing new infectious virus particles.

The rate of disease progression depends on several variables, including the health of the host and the particular strain of the virus. However, the median time from initial infection to the development of AIDS among untreated patients is typically 8 to 10 years.

HIV damages the body's sources of T cells and centers of immune activity, killing cells in the bone marrow and thymus that are needed for the development of mature immune cells. The virus also progressively destroys the lymph nodes, the centers of immune activity in the body. The viral load, or amount of virus in circulation, is a barometer of disease progression.

Infectious Knowledge

HIV infection is not like catching a cold or the flu, because it isn't spread by coughs or sneezes. You get HIV by coming in contact with infected blood, semen, or vaginal fluids from another person.

Fiction:

- ◆ HIV can be spread by everyday regular contact with infected people at school, work, and home.
- ◆ HIV infection can occur from contact with infected clothes, phones, toilet seats, or sharing eating utensils.
- ◆ HIV can be spread through a blood transfusion from a regulated blood bank.
- ◆ Mosquitoes transmit HIV.
- ◆ HIV can be acquired from sweat, saliva, or tears.
- ◆ Any type of kissing will spread HIV.

Fact:

- ◆ Sharing needles for injectable drugs can transmit HIV.
- ◆ Unprotected sex, whether with males or females, can spread HIV.
- ◆ Infection with other sexually transmitted diseases increases the risk of HIV infection.
- ◆ Unscreened blood transfusions or blood products from high-risk individuals can spread HIV.
- ◆ HIV-positive mothers may transmit HIV through breast milk.

Opportunistic Infections

The relative risk of developing opportunistic infections is closely tied to T cell counts. As T cell counts go down, many different organisms can cause infections. Common AIDS-related opportunistic infections are caused by fungi like Candida (topical and systemic yeast infections), bacteria like tuberculosis, and viruses like hepatitis and herpes.

HIV Symptoms

Early symptoms of HIV infection include …

- ◆ Swollen glands
- ◆ Night sweats or continuous fever
- ◆ Unexplained weight loss exceeding 10 pounds

◆ Heavy, continual dry cough

◆ Shortness of breath

◆ Oral thrush

◆ Recurrent vaginal yeast infections

◆ Unexplained skin rashes

◆ Herpes infections that require a long time to heal

A common physical manifestation of late-stage AIDS is the "Wasting Syndrome," when body fat and body mass are significantly decreased, making a person look like they are suffering from severe malnutrition.

Where Did HIV Come From?

The best accepted theory about the origin of HIV is that it is a descendant of a closely related virus, simian immunodeficiency virus (SIV), which infects monkeys. Researchers have known for a long time that certain viruses can pass from animals to humans, a process that is called *zoonosis*. HIV may have crossed over from chimpanzees as a result of a human killing a chimp and eating it for food.

Disease Diction

A **zoonosis** is an infection or infectious disease that is transmissible from vertebrate animals to people.

Potent Fact

Why don't mosquitoes transmit AIDS? The HIV virus does not multiply in mosquitoes. If a mosquito feeds on an HIV-infected human, the virus is treated like food and digested along with the blood meal. If the mosquito resumes feeding on a non-HIV-infected individual, too few particles are transferred to initiate a new infection.

In 1999, researchers confirmed that tissue from a chimpanzee carried a form of SIV that was nearly identical to an aggressive form of HIV, HIV-1. It appears highly likely chimpanzees were the source of HIV-1, and that the virus at some point crossed species from chimpanzees to humans. We cannot say for sure when the virus first emerged, but it is clear that HIV started to infect humans and became epidemic in the middle of the twentieth century.

There are a number of factors that allowed HIV to move from epidemic to pandemic, but international travel was a key factor. The first person diagnosed with AIDS, and potentially an early source of the HIV epidemic in the United States, called "patient zero," was a Canadian flight attendant named Gaetan Dugas. He traveled extensively, and analysis of several early cases of AIDS showed that these individuals had either direct or indirect (shared partner) sexual contact with the flight attendant. The early cases were also traced to several different American cities where these infected individuals lived, demonstrating a critical role for air travel in spreading the virus.

Hooper's Theory?

Not long ago, British author Edward Hooper claimed that HIV originated as a product of cross-contamination from an oral polio vaccine administered in Africa in the late 1950s. The vaccine, called "Chat," was claimed to be derived from chimp kidney cells and was thought to be contaminated with SIV. The vaccine was given to roughly a million people in the Belgian Congo, Rwanda, and Burundi.

On the surface, Hooper's claim seemed to have merit, because the oldest known case of AIDS occurred in a man from Kinshasha near the mouth of the Congo River. His blood was drawn in 1959 and tested positive for HIV. However, the Wistar Institute in Philadelphia, which developed the polio vaccine, came across an old vial containing the original vaccine. After careful analysis, it was proven that the vaccine contained neither HIV nor SIV. In addition, they confirmed that only macaque monkey kidney cells, which cannot be infected with SIV or HIV, were used to make the Chat vaccine. Bottom line, always check your facts, and think twice before throwing old stuff away.

First Human HIV Infection

A plasma sample taken in 1959 from an adult male living in the Democratic Republic of Congo was found to contain HIV. The virus was found in tissue samples from an African American teenager who died in St. Louis in 1969, and HIV was found in tissue samples from a Norwegian sailor who died around 1976. Many scientists believe that HIV was probably introduced into humans around the 1940s or the early 1950s, although it has been suggested that the first case may actually have occurred in the 1930s.

Infectious Knowledge

Actor Rock Hudson was the first major public figure known to have died of AIDS. Tennis great Arthur Ashe also died from transfusion-related AIDS. Ryan White, a 13-year-old hemophiliac with AIDS, who eventually died, became famous for his fight to attend school. And basketball superstar Magic Johnson, 10 years after he announced that he was HIV positive, is still living a healthy life with the disease.

Early Anti-HIV Therapy

In 1987, AZT (zidovudine, Retrovir) became the first anti-HIV drug approved by the Food and Drug Administration (FDA). It blocks the activity of the critical HIV enzyme needed to make new virus particles, restricting the virus's ability to replicate. Unlike bacteriocidal antibiotics, which can kill certain infectious bacteria, anti-HIV therapy is bacteriostatic, and only limits production of new virus—it doesn't kill the virus that's already there.

AZT therapy helped HIV-infected patients live longer with fewer opportunistic infections because it helped to reduce viral load. However, drug resistance and side effects limited AZT therapy. Common side effects included nausea, vomiting, headache, fatigue, weakness, and/or muscle pain. Other side effects included inflammation, insomnia, and kidney disorders. Possibly the best application of AZT was its use during pregnancy as a way to reduce the risk of HIV being passed from mother to child.

> **Infectious Knowledge**
>
> Effective therapy against HIV is complex and requires that patients take several different drugs at very specific times. Some have to be taken with food, some on an empty stomach. This is one of the reasons why it is difficult to deliver effective therapy in undeveloped countries.

You've Got to Have HAART

New approaches were needed to advance HIV therapy. In the mid-1990s, there was a significant breakthrough in therapy with the approval of a new class of drugs called "protease inhibitors." These drugs are highly active in blocking HIV replication when used alone. Yet HIV's ability to mutate and become resistant threatened to make these new drugs useless. When treating an infection caused by a rapidly mutating virus like HIV, the most effective method is combination therapy with multiple drugs that act in different ways. This way, a single mutation cannot cause resistance.

Today there are a number of *antiretroviral* agents available as therapeutic options that have several different mechanisms of action.

> **Disease Diction**
>
> **Antiretroviral** describes a drug that acts against a retrovirus such as HIV.

Highly active antiretroviral therapy (HAART) consists of three-drug combinations of antiretroviral agents. This approach has significantly decreased the incidence of death among HIV-infected people in the United States and has allowed HIV-infected individuals to lead productive lives. Decreasing amounts of HIV in the bloodstream helps restore immune function, so opportunistic infections are less of a problem. During therapy, it's important to measure viral load and T cell counts to monitor the status of HIV disease and to guide recommendations for continued therapy.

But HAART, like all complex medications, has problems. There are serious long-term side effects, such as an accumulation of lactic acid in the bloodstream and physical and metabolic changes that cause changes in fat distribution as well as cholesterol and glucose abnormalities that can lead to a risk of heart disease. Long-term use of the drugs can also promote the development of drug-resistant strains of HIV. Clearly, these problems must be balanced with the alternative of a greatly shortened life. Finally, HAART therapy is expensive and requires strict compliance for proper administration. For this reason, it is not a realistic option for many of the 40 million HIV-infected people in the world.

Diagnosing HIV Infection

A positive diagnosis of HIV infection can be made with several types of tests, including virus culture, antibody testing, and polymerase chain reaction (PCR) (see Chapter 5 for more on PCR). Generally, HIV infection is diagnosed by showing that specific antibodies fighting the infection are present in the blood. Antibodies against HIV are usually detectable approximately 25 days after infection, and nearly all infected individuals are HIV antibody-positive after 12 weeks.

False-negative tests can arise because it may take several weeks after initial infection for the body to produce antibodies. Rapid tests can generate results in less than 30 minutes, analyzing blood, saliva, or urine.

> **CAUTION**
>
> **Antigen Alert**
>
> One of the major reasons antiretroviral treatment of HIV infections has not been more successful is the development of drug-resistant HIV strains in patients during therapy. Identifying such strains early can help improve the effectiveness of treatment.

HIV Vaccine: The Holy Grail of Containment

In the mid-1980s, United States Health and Human Services Secretary Margaret Heckler predicted a brief epidemic for AIDS and said, "there will be a vaccine in a very few years and a cure for AIDS before 1990." Today, it is still recognized that the best hope for stopping the pandemic lies with a vaccine. Vaccines eradicated smallpox and contained polio, but the prospect for an AIDS vaccine is nowhere in sight.

How Can This Be?

One problem is that HIV protects itself by producing chemicals that block antibodies from killing it. However, the larger problem relates to the number of different strains of HIV. In reality, HIV is not a single virus, but a family of related virus subtypes called "clades." The AIDS pandemic has been caused by eight clades, labeled A through H, of HIV-1. In addition, new strains are continuously being generated because HIV changes rapidly through mutation and genetic recombination, which mixes genetic information from virus subtypes to create new ones.

Potent Fact

An AIDS vaccine that works in the United States may not be effective in other parts of the world because there are different strains that are predominant in different geographic areas.

The eight HIV-1 subtypes are not evenly distributed around the world. Rather, they exist in distinct geographic clusters. For example, subtype B is most prevalent in Europe, North and South America, Japan, and Australia, while subtypes A, C, D, and E are spreading rapidly in Africa and Asia. This family diversity suggests that a single vaccine would be unlikely to cover all clades. In fact, some researchers lament that an effective vaccine for industrialized countries would not work against the HIV subtypes prevalent in the developing world.

Researchers generally agree that for a vaccine to be truly effective both humoral (antibody) and cell-mediated (B cells and killer T cells) arms of the immune system need to be activated. Antibodies can protect against initial infection, while B cells and killer T cells confer long-term immunity. Regrettably, our current vaccine candidates have been unable to induce both types of protection.

Yet the search continues with glimmers of hope. Researchers have made progress in prodding the body to track down and destroy HIV-infected cells for up to two years. Even if a vaccine fails to prevent infection, such medications could still limit the course of disease, keeping viral loads low, and reduce the risk of serious complications.

Natural Resistance to HIV Infection

Some people with HIV exposure don't progress to AIDS. In the mid to late 1990s, scientists noted that persons with one copy of a mutated gene were less likely to become infected with HIV, while individuals inheriting two copies of the mutated gene were highly resistant to HIV infection. The mutated gene, called CCR5-32, helps HIV invade T cells. The gene defect prevents the host cell from binding with the HIV. Unfortunately, the gene defect is rare, with about 1 percent of whites having two genes with the CCR5 mutation and even fewer nonwhites.

Lessons Learned

Perhaps the most important lesson of the past 20 years is that the HIV/AIDS pandemic has continued to escalate despite our greatest technological efforts. We've known the entire genetic code of HIV for more than 15 years, and we understand its biology. But humbly, we really don't understand enough—especially the intricacies of the way HIV interacts with the human body. HIV is simple by appearance, yet complex beyond our wildest dreams. A vaccine and a cure remain elusive.

The sad fact is that millions will die each year and the pandemic will continue to escalate. However, as a world community, we will persevere. Combined antiviral therapy promises longer and higher-quality survival for HIV-infected people, and cheaper regimens makes it likely that more of the world's infected population will live longer. Ultimately, research is our best hope.

The Least You Need to Know

- HIV is a dangerous infectious disease that is spread primarily through sexual contact.
- Practicing safe sex—particularly using condoms—helps to prevent HIV transmission.
- Although AIDS is not curable, in the United States it can be controlled through antiretroviral therapy, although there can be problems with side effects from the many drugs that must be taken.
- In certain parts of the world, AIDS is much more common and problematic than it is in the United States.
- An AIDS vaccine would be tremendously helpful, but it is difficult to make one that works because the virus changes so quickly and because there are several different strains of HIV.

A Disease Without Boundaries: Tuberculosis

In This Chapter

- ◆ Causes and symptoms of tuberculosis
- ◆ Treating tuberculosis
- ◆ Importance of good public health
- ◆ Multi-drug resistant strains

In an age when we believe that we have the tools to conquer most diseases, the ancient scourge of tuberculosis (TB) still causes 2 million deaths a year worldwide—more than any other single infectious organism—reminding us that we still have a long way to go. Even equipped with drugs to treat TB effectively, we haven't managed to eradicate this deadly infection.

What is the history of tuberculosis? And how has it managed to survive for so long? This chapter will answer these questions, plus describe the symptoms and treatment options available for TB.

An Ancient Scourge That Still Kills Today

Mycobacterium tuberculosis, the bacteria that causes *tuberculosis*, has been around for centuries. Recently, fragments of the spinal columns from Egyptian mummies from 2400 B.C.E. were found to have definite signs of the ravages of this terrible disease. Also called consumption, TB was identified as the most widespread disease in ancient Greece, where it was almost always fatal. But it wasn't until centuries later that the first descriptions of the disease began to appear. Starting in the late seventeenth century, physicians began to identify changes in the lungs common in all consumptive, or TB, patients. At the same time, the earliest references to the fact that the disease was infectious began to appear.

In 1720, the English doctor Benjamin Marten was the first to state that TB could be caused by "wonderfully minute living creatures." He went further to say that it was likely that ongoing contact with a consumptive patient could cause a healthy person to get sick. Although Marten's findings didn't help to cure TB, they did help people to better understand the disease.

Disease Diction

Tuberculosis was first formally described by Greek physician Hippocrates around 460 B.C.E. He called it *phthsis*, which is the Greek word for consumption, because it described the way the disease consumed its victims. Consumption was the most widespread disease of the time, and most of its victims died. The word consumption was used to describe the disease until 1882, when the *tuberculosis* bacteria was identified as the cause of the disease.

The sanitorium, which was introduced in the mid-nineteenth century, was the first positive step to contain TB. Hermann Brehmer, a Silesian botany student who had TB, was told by his doctor to find a healthy climate. He moved to the Himalayas and continued his studies. He survived his bout with the illness, and after he received his doctorate, built an institution in Gorbersdorf, where TB patients could come to recuperate. They received good nutrition and were outside in fresh air most of the day. This became the model for the development of sanitoria around the world.

Antigen Alert

Tuberculosis is spread through the air, so everyone is at some risk.

In 1865, French military doctor Jean-Antoine Villemin demonstrated that TB could be passed from people to cattle and from cattle to rabbits. In 1882, Robert Koch discovered a staining technique that allowed him to see the bacteria that cause TB under a microscope.

Until the introduction of surgical techniques to remove infected tissue and the development of x-rays to monitor the disease, doctors didn't have great tools to treat TB. TB patients could be isolated, which helped reduce the spread of the disease, but treating it remained a challenge.

Tuberculosis: Airborne Nightmare

TB-causing bacteria is passed from person to person through the air when someone with the disease coughs or sneezes. People who are nearby may get infected after breathing in bacteria. The bacteria can attack any part of the body, but they usually stick to the lungs.

People with TB disease are most likely to spread it to those they spend time with every day, like their family or co-workers.

Latent TB Infection

Only 5 to 10 percent of healthy people who come in contact with TB bacteria will ever get sick. The vast majority of them will live with dormant TB bacteria in their bodies throughout their lives, because their immune systems are able to fight the bacteria and stop them from growing. People with latent TB don't feel sick, don't have symptoms, and can't spread TB. However, the bacteria remain alive in the body and can become active later. They have what are called latent infections.

If at some point in their lives their immune system is weakened, the once-dormant bacteria may begin to grow again and cause active tuberculosis. Sometimes, doctors will recommend that people with latent TB infections take medicine to prevent development of active disease. The medicine is usually a drug called isoniazid (INH), which kills the TB bacteria that are in the body. Usually the course of treatment is six to nine months. Children and people with HIV infection, however, sometimes have to take INH for a longer period of time.

Potent Fact

Pulmonary tuberculosis, or TB of the lungs, is the most common form of TB. TB can also attack the spine, bones and joints, the central nervous system, the gastrointestinal tract, the lymph system, and the heart.

Active TB Disease

TB bacteria become active if the body's immune system can't stop them from growing. They then multiply and make people sick. A small number of people get sick soon after they are infected, but most reactivate after years of latent infection.

Babies, young children, and people infected with HIV have weak immune systems and are more likely to develop active TB. Other conditions, like diabetes, leukemia, severe kidney disease, low body weight, and substance abuse, also can make a person more likely to come down with an active case of TB.

Symptoms of TB depend on where in the body the bacteria grow, but most of the time they grow in the lungs. When they do, the symptoms are …

Potent Fact

TB is spread through the air, not through handshakes, sitting on toilet seats, or sharing dishes and utensils with someone who has TB. However, casual exposure is not sufficient for someone to get TB.

- A bad cough that lasts longer than two weeks.
- Pain in the chest.
- Coughing up blood or sputum.
- Weakness or fatigue.
- Weight loss.
- Loss of appetite.
- Chills.
- Fever.
- Sweating at night.

Disease Diction

Latent TB infection   A person with a latent TB infection has no symptoms and does not feel sick. They cannot spread TB to others although if they are tested for it, the test will indicate that they have been exposed to the bacteria. Chest x-rays and sputum tests will be negative. In latent infection, the immune system "walls off" the bacteria, which form a thick, waxy coat and can lie dormant that way for years.

Active TB infection   A person with active TB has symptoms, including cough, chest pain, coughing up blood or sputum, weakness, fatigue, loss of appetite, chills, fever, and night sweats. They can spread TB to others, and a skin test for the disease will show positive results. They may also have an abnormal chest x-ray and/or positive sputum smear or culture.

Active TB is diagnosed by taking a sputum sample and seeing if TB bacteria grow in a culture. X-rays may also be used to show the presence of bacteria in the lungs.

Left untreated, a person with active TB will infect an average of between 10 to 15 people per year.

Treating TB

Most of the time TB can be cured with antibiotics. If you have TB, you will need to take several drugs. This is because there are many bacteria to be killed. Taking multiple drugs also helps to prevent the bacteria from becoming drug resistant and, thus, much more difficult to cure.

If you have TB of the lungs, or pulmonary TB, you are probably infectious. This means that you can spread the disease by coughing or sneezing. Fortunately, after a couple of weeks of taking medicine, most people are no longer infectious and they begin to feel better. Usually they can return to life as usual. But that doesn't mean all the bacteria are killed. People often have to take TB medicine for six to nine months before all the bacteria are killed.

Why Is It Important to Take TB Medicine for So Long?

TB bacteria die very slowly. Even when patients start to feel better, the bacteria are alive in their bodies. They have to keep taking medicine until all the bacteria are dead, otherwise they can get sick again and infect others.

Another danger of not completing the whole course of therapy is the rise of drug-resistant TB. If you stop taking your medicine and some of the bugs are still alive, they may become resistant to the drugs you were taking, so that if you get sick again, you will need different drugs to kill the bacteria because the old ones won't work. These additional drugs, called second-line drugs, must be taken for a very long time, sometimes up to two years, and their side effects can be quite serious.

The only way to get better is to take your medicine as prescribed by the doctor. Most public health officials advocate Directly Observed Therapy (DOTS), which is when a health-care worker meets with the patient every day, or several times a week, to be sure they take their medicine. Sometimes they meet at the patient's home or at a hospital or TB clinic. Some DOTS programs provide medicine that can be taken only two or three times a week instead of every day. In addition to ensuring that the patient takes their medication as prescribed, the health care worker also monitors side effects.

Potent Fact

Cure rates for TB are above 90 percent if medicine is taken properly and to completion. For patients with multi-drug resistant TB, cure rates are only 50 percent.

Infectious Knowledge

Statistics have shown that income level has nothing to do with who takes their TB medicine to completion. Doctors and wealthy people are just as likely to stop taking their medicine as those who are less fortunate.

DOTS works and it is used in many countries. It is the World Health Organization's recommended method for successfully treating TB.

Patients with active TB who have to go to the hospital may be put in special rooms with negative air pressure. This keeps TB from spreading from room to room, or out into hospital hallways. People who enter the rooms will wear special facemasks to protect themselves.

Multi-Drug-Resistant TB (MDR-TB)

When TB patients don't take their medicine properly, the TB bacteria may become resistant to certain drugs. This means the drugs can no longer kill the bacteria. Drug resistance is most likely to occur when people ...

◆ Have spent time with someone with drug-resistant TB.

◆ Don't take their medicine regularly.

◆ Don't take all their medicine.

◆ Develop TB after they've taken TB drugs before.

◆ Come from areas where drug-resistant TB is common.

Multi-drug-resistant TB (MDR-TB) is a form of tuberculosis that is resistant to two or more of the first-line drugs used to treat the disease. When the bacteria resist the antibiotics used to attack them, they relay that ability to new bacteria that is produced. People with multi-drug-resistant TB must be treated with special second-line drugs. These drugs don't kill the bacteria as well as the first-line drugs, and they often cause more severe side effects.

Infectious Knowledge
DNA fingerprinting of TB strains can track the spread of a particular strain, confirm that an outbreak has occurred, and show the difference between cases in which a latent infection has been reactivated from those that are caused by recent infection.

If a person with MDR-TB spreads the disease to someone else and that person comes down with active disease, it will be multi-drug-resistant from the beginning. In the early 1990s, there were several outbreaks of multi-drug-resistant TB in New York City hospitals that were caused primarily by the spread of one strain, strain W, that went from patient to patient to patient. This strain was resistant to between seven and nine drugs. A large number of these patients died, and many health-care workers now have latent infections with this highly resistant strain.

The success in treating MDR-TB depends on how quickly it is identified and whether effective drugs can be found. Unfortunately, tests to determine whether a particular strain is resistant usually take several weeks to complete. During the delay, the patient may be improperly treated and therefore remain infectious. In populations at high risk for MDR-TB, this danger is combated by starting treatment without waiting for susceptibility results to confirm which drugs will be most effective.

TB is difficult to diagnose and drug resistance takes a long time to determine because the bacteria grows so slowly. Most bacteria double in 20 minutes—TB takes 24 hours. So it takes 24 hours to get 2 bacteria, 48 hours to get 4, and so on.

Treatment for MDR-TB can take up to two years and cost up to $250,000. The cure rate is only 50 percent, while the vast majority of drug-sensitive TB is curable. The World Health Organization estimates that over 50 million people worldwide are infected with MDR-TB. Given the difficulty of treating MDR-TB, this is a very frightening statistic.

Infectious Knowledge
A strain of MDR-TB originally develops when a case of drug-susceptible TB is improperly or incompletely treated. This occurs when a doctor doesn't prescribe the right drugs or when a patient doesn't take the drugs properly. This allows individual bacteria that have natural resistance to a drug to multiply. Over time, the majority of bacteria in the body become resistant.

TB Is a Global Peril

Despite the existence of effective drugs, each year more people die of TB. Recent outbreaks have occurred in Eastern Europe, where TB deaths are increasing after many years of steady decline. The largest number of cases is in Southeast Asia.

Each year approximately eight million people get active TB. More than 1.5 million of those cases are in sub-Saharan Africa. This number is rising rapidly because of the HIV/AIDS epidemic, which makes TB easier to catch because it suppresses the body's immune defenses. Nearly three million TB cases each year are in Southeast Asia. Russian prisons have extremely high TB rates and many inmates have the drug-resistant strain. This becomes an even greater health risk when prison amnesty is granted to tens of thousands at a time and the infected prisoners are released to cause disease among civilians.

Potent Fact

One third of the population of the world is infected with TB. The vast majority of those have latent infections.

Factors Contributing to the Rise of TB

A number of factors have contributed to the rise of tuberculosis in recent years. They include the scary combination of HIV and TB infection, a rapid increase in global travel, malnutrition, overcrowding, and an increase in the number of refugees who carry the disease.

Antigen Alert

A healthy person infected with TB has a 5–10 percent chance of getting active disease in his or her lifetime. An HIV-positive person has a 5–10 percent chance of getting active TB *each year*.

HIV Accelerates the Spread of TB

HIV and TB are a lethal combination, with each one speeding the others' progress. HIV weakens the immune system so that if an HIV-positive person becomes infected with TB, they are much more likely to develop an active case of the disease. As a result, TB causes 15 percent of HIV deaths worldwide. In Africa, HIV is the most important factor that has led to the increased incidence of TB in the past 10 years.

Poor Management

Poor management of TB programs threatens to make a curable disease incurable.

Today there are resistant TB strains in every country that has been surveyed, with some strains resistant to all major anti-TB drugs. In undeveloped nations, doctors sometimes prescribe the wrong medicine, don't treat with multiple drugs at one time, or their drug supply is unreliable. The rise of MDR-TB, especially in the former Soviet Union, threatens global TB control efforts. And with the cost of treating MDR-TB so prohibitive in many countries, major outbreaks are possible.

Potent Fact

Can you get TB on a plane?

Yes. It is not likely, but it is possible. There are documented cases of TB infection that have occurred on long international flights. The people exposed to and infected with TB were those who sat closest to the person with the active disease.

Public health officials worry that poorly supervised or incomplete treatment of TB is sometimes worse than no treatment at all. Unlike those who are treated properly and stop being infectious after a few weeks (although they aren't cured yet), people who take the wrong medicines, or don't take them for long enough, are likely to remain infectious and spread their disease to many others. Drug resistance can arise and spread through improper and/or poorly supervised treatment as well.

Global trade and travel has increased dramatically over the past 40 years, helping to spread TB and other infectious diseases. There are proven cases of people infecting others on long plane rides and cases of MDR-TB strains that have been carried by a patient from one country to another.

Refugees and Displaced People

Another factor hastening the spread of TB is the rising number of refugees and displaced people in the world. Untreated TB spreads rapidly in crowded camps and shelters, where it is estimated that up to 50 percent of refugees may be infected. Displaced people are a danger because it is difficult to treat people who are constantly moving around; as they move, they spread the disease.

Other displaced people, like the homeless in industrialized countries, are also at risk. In 1995, close to 30 percent of San Francisco's homeless population and 25 percent of London's homeless were reported to be infected with TB. These infection rates are much higher than the rates in the general population.

Although tuberculosis is a curable disease, it continues to cause much disease and death worldwide. Without aggressive public health measures and continued research on effective treatments and vaccine development, this treatable disease will continue to run rampant.

The Least You Need to Know

- ◆ Tuberculosis is caused by a bacteria that is spread through the air. Everyone is at some risk of developing a TB infection.
- ◆ The vast majority of healthy people with TB have latent infections and don't get sick. A small percentage will get sick if their immune system is temporarily or permanently compromised.
- ◆ Only people with active TB disease can spread the disease to others.
- ◆ Treatment for TB involves several drugs and takes six to eight months. Unless treatment is completed, the bugs aren't all killed and you can get sick again, often with a drug-resistant strain.
- ◆ MDR-TB, the drug-resistant strain of TB, is difficult and expensive to treat and can only be cured about 50 percent of the time.

Chapter

Still Going Strong After All These Years: Malaria

In This Chapter

- ◆ History and evolution of malaria as a disease
- ◆ Life cycle of malaria from mosquito to human
- ◆ Diagnosis and treatment
- ◆ Developing a vaccine

Malaria is one of the most successful parasites ever known to mankind. After thousands of years, it remains the world's most pervasive infection, affecting at least 91 different countries and some 300 million people. The disease causes fever, shivering, joint pain, headache, and vomiting. In severe cases, patients can have jaundice, kidney failure, and anemia, and can lapse into a coma.

It is ever-present in the tropics and countries in sub-Saharan Africa, which account for nearly 90 percent of all malaria cases. The majority of the remaining cases are clustered in India, Brazil, Afghanistan, Sri Lanka, Thailand, Indonesia, Vietnam, Cambodia, and China. Malaria causes 1 to 1.5 million deaths each year, and in Africa, it accounts for 25 percent of all deaths of children under the age of five.

A Brief History of Malaria

Ancient accounts of malaria date back to Vedic writings of 1600 B.C.E. in India and to the fifth century B.C.E. in Greece, when the great Greek physician Hippocrates, often called "the Father of Medicine," described the characteristics of the disease and related them to seasons and location. The discovery of an association of malaria with stagnant waters led the Romans to develop drainage programs, which were among the first documented preventions against malaria. In seventh-century Italy, the disease was prevalent in foul-smelling swamps near Rome and was named *mal' aria* Italian for "bad air."

Infectious Knowledge

Some historians believe that malaria epidemics greatly contributed to the fall of the Roman Empire. DNA from the 1,500-year-old bones of a child found in a cemetery near Rome yielded evidence of a malaria epidemic. A large epidemic may explain why one of the greatest military machines in world history was too weak to repel invasions from the Visigoths, Huns, and Vandals. Did moral and urban decay befall Rome, or was it malaria?

Malaria epidemics ravaged Europe and Africa for centuries. Like many diseases, it traveled with tradesmen, settlers, and conquering forces. Over four centuries of the slave trade, millions of Africans died from malaria, which may have come to the New World along with slaves.

Despite malaria's preference for the tropics, the disease has had an impact on the history of the United States, too. Known commonly as "fever and ague," malaria took its toll on early American settlers. In the book *Little House on the Prairie*, Laura Ingalls Wilder vividly describes its impact. Malaria devastated the 1607 Jamestown colony and regularly ravaged the South and Midwest. The incidence of malaria in the United States peaked in 1875. Yet, in 1914, more than 600,000 new cases were still occurring.

Infectious Knowledge

Sir Ronald Ross, born in India in 1857, received the 1902 Nobel Prize in Medicine for his pioneering work on malaria, in which he laid the foundation for successful research on this disease and methods for combating it.

Malaria has been a factor in nearly all United States military campaigns. During the Civil War, armies on both sides sustained more than 1.2 million cases of malaria. It continued to be a problem in both World Wars, the Korean War, and the Vietnam War. In the latter, malaria appeared in a newer, more deadly, drug-resistant form.

A Nasty Parasite

Charles Lavern, a French army surgeon in Algeria in 1880, first described malaria parasites in the red blood cells of humans. Several years later, Sir Ronald Ross observed developing parasites in the intestines of mosquitoes and provided the first major evidence that mosquitoes were acting as *vectors*, or vehicles, to spread disease. Ross succeeded in demonstrating the life cycle of the parasites of malaria in mosquitoes.

Human malaria is caused by four main species of the *Plasmodium* parasite:

- *P. falciparum:* the most important of the malaria parasites because it can be rapidly fatal and is responsible for the majority of malaria-related deaths. It predominates in Africa, New Guinea, and Haiti.

- *P. malariae:* found occasionally in *endemic* areas like sub-Saharan Africa.

- *P. vivax:* more common on the Indian sub-continent and Central America, with the prevalence of these two infections roughly equal in Asia, Oceania, and South America.

- *P. ovale:* mostly confined to Africa, although sporadic cases occur in Southern India.

Disease Diction

A **vector** is a vehicle for moving a disease-causing organism from one host to another. Mosquitoes were vectors for malaria by helping to spread it from human to human.

Disease Diction

A disease is **endemic** when it is constantly present in a community or among a group of people.

Transmission: Mosquito Marauder and a Deadly Cycle

Malaria parasites are transmitted by the female anopheles mosquito. There are about 380 species of this type of mosquito, but only about 60 species can transmit disease. The parasites can only live within female mosquitoes and can be transferred only to humans. The parasites have a complex life cycle that is split between the human host and the mosquito vector. The process of malaria transmission occurs this way:

Antigen Alert

Malaria kills one child every 30 seconds. A total of 800,000 children under the age of five die from malaria every year, making this disease one of the major causes of infant and juvenile mortality. The disease is also responsible for a substantial number of miscarriages and low-birth-weight babies.

The life cycle of the malaria parasite.

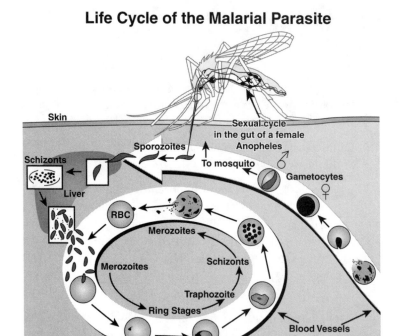

Life Cycle of the Malarial Parasite

1. A female mosquito harboring malaria feeds on human blood and transmits thread-like structures, called *sporozoites*, to the human.

2. The sporozoites travel to the liver and multiply. They mature over two to four weeks without causing disease symptoms.

3. The mature sporozoites, called merozoites, are released into the bloodstream, where they penetrate red blood cells and multiply and break down hemoglobin, which is essential for oxygen transport.

4. The blood cells degrade, and the merozoites escape and infect other blood cells. This induces bouts of fever, chills, sweating, and anemia in the infected individual. The infected red cells can obstruct blood vessels in the brain (which is called cerebral malaria) or other vital organs, leading to the death of the patient.

5. A few parasites form a sexual stage, which can be sucked up by another mosquito taking a blood meal, beginning a new transmission cycle.

Disease Diction

A **sporozoite** is a slender, spindle-shaped organism that is the infective stage of the malaria parasite. It is the result of the sexual reproductive cycle of the parasite, which occurs inside the mosquito.

6. Two sexually active parasites meet in the mosquito's gut and produce a new generation.

The parasites must spend about two weeks in the mosquito to undergo further life cycle changes before they can infect humans again. When the mosquito feeds on another human, the parasites are injected into a new host. This mosquito can transmit the infection only if she sucks more blood from an uninfected person before she dies.

Fever and Ague: Malaria's Symptoms

People infected with malaria typically experience fever, shivering, pain in the joints, headache, vomiting, convulsions, and coma. Malaria is especially dangerous to pregnant women and small children. Different malarias produce fevers of varying frequency. Severe *anemia* is often the cause of death in endemic areas.

If not treated, the disease progresses to severe malaria. Malaria is especially dangerous to pregnant women and small children. Severe malaria results in coma, jaundice, kidney failure, severe anemia, and/or high parasite counts and should be treated in an intensive care unit where patients can be monitored closely.

Potent Fact

Malaria is not common in the United States, but people who travel to parts of the world where it is endemic need to protect themselves by taking antimalarial medication and trying to avoid mosquito bites by using repellent, netting, and so on.

Disease Diction

Anemia is a condition when a person doesn't have enough red blood cells, or hemoglobin, in the body. If hemoglobin is low, the blood can't carry enough oxygen to organs. Symptoms of anemia include weakness, faintness, shortness of breath, increased heart beat, headaches, sore tongue, nausea, loss of appetite, dizziness, bleeding gums, yellow eyes and skin, confusion, and dementia. Severe cases may have signs of heart failure.

Sickle Cell Anemia: Natural Resistance at a Price

Sickle cell anemia is a hereditary disease that affects molecules in the blood, resulting in sickle-shaped red blood cells. Individuals carrying one normal and one sick gene are called "heterozygous" carriers of the disease. They produce both normal and sickle-shaped blood cells because neither gene is dominant. It has been observed that heterozygous carriers of the disease rarely develop malaria.

It appears that sickle cell anemia occurs frequently in African populations as a result of earlier generations' exposure to malaria. In Central Africa, where malaria has been epidemic for a very long time, nearly 45 percent of the population (45 people out of 100) carry one gene for the sickle-cell trait that confers resistance to malaria. In the United States, 10 percent of African Americans are sickle-cell carriers. The presence of malaria worldwide continues the selection process and maintains the heterozygous sickle-cell population.

Disease Diction

Sickle cell anemia is a hereditary blood disorder that affects hemoglobin. Sickle-shaped cells clog blood vessels and keep oxygen from getting to the body's tissues and organs. Symptoms include pain in chest, stomach, arms, legs, and bones, jaundice, tiredness, slow growth in children and late onset of puberty. There is no cure for sickle cell anemia. Treatment may include vitamins, prevention of dehydration, and blood transfusions.

Diagnosing Malaria

Malaria is diagnosed by its symptoms and microscopic examination of blood to identify the parasites in different stages of development. Ideally, blood should be collected when the patient's temperature is rising, as that is when the greatest number of parasites is likely to be found. Three consecutive days of tests that do not indicate the presence of the parasite can rule out malaria.

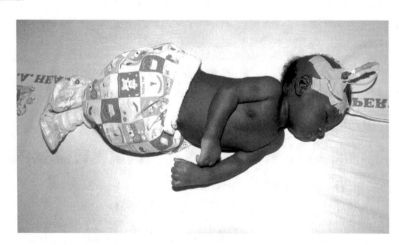

Baby in advanced stage of malaria at Garki General Hospital in Abuja, Nigeria.

(© WHO/Pierre Viyot)

The presence of antibodies in the blood can help a physician determine whether a patient has been exposed to malaria, but it does not differentiate between present and past infections. In other words, the antibodies might have been present from a previous malaria infection that has since been cured, and so the presence of antibodies may indicate that a patient is reinfected, but it may not.

Rapid diagnostic tests for malaria are now available that allow the detection of antigens in a finger-prick blood sample in minutes, with sensitivity similar to that achieved by examination under a microscope. Easy-to-use dipstick tests that detect specific proteins and enzymes have the potential to enhance the speed and accuracy of diagnosis, particularly when tests are performed by untrained individuals.

A Frustrating, Yet Curable, Disease

Malaria is a curable disease if diagnosed promptly and adequately treated. In China, an infusion of the qinghao plant has been used for at least the last 2,000 years to relieve malaria symptoms. The bitter bark of the cinchona tree was used in Peru before the fifteenth century for the same purpose. Quinine, the tree's primary active ingredient, was isolated in 1820. For nearly 300 years and until the 1930s, quinine was the only effective agent for the treatment of malaria. But it is now only used for treating severe malaria because of undesirable side effects. Common side effects include stomach cramps, nausea, diarrhea, and vomiting. Less common side effects can be dizziness, ringing in the ears, skin rash, and visual disturbances.

Infectious Knowledge

A gin and tonic a day keeps malaria away. Not quite. The popular alcoholic mix contains 20 milligrams of quinine per six fluid ounces. A typical daily dose of quinine for malaria treatment is more than 1000 milligrams per day. However, party-goers can thank malaria for the origin of the "Gin and Tonic," since to make quinine's bitter taste more palatable, British colonials in India mixed it with gin and lemon or lime. Tonic water was granted an English patent in 1858.

Various alternatives to quinine have been used since the 1930s, yet they each had negative side effects that limited their application, or the bacteria developed strains that were resistant to the treatment. Today, drug-sensitive malaria is largely controlled by intravenous (IV) chloroquine, while drug-resistant forms are treated with quinine or quinine-derivatives combined with antibiotics.

The emergence of resistance to widely used antimalarial drugs such as chloroquine has greatly limited the control and treatment of malaria in certain countries. The extensive use and misuse of antimalarial drugs has contributed to the spread of resistance. It is important for drugs to be used properly, and in combination when necessary. New drugs would be helpful in this area, too.

Malaria's Many Costs

The economic burden imposed by malaria is enormous—the disease strains limited health-care resources and diminishes productivity.

In endemic areas, 3 in 10 hospital beds are occupied by victims of malaria. Seasonal malarial outbreaks can devastate farms as they frequently coincide with the rainy season and harvest time.

World map showing areas (shaded) where malaria is ever-present, or endemic.

(World Health Organization, 1997)

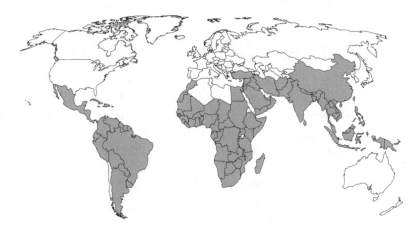

Potent Fact

Global warming and other climatic events, such as El Niño, help spread malaria to different geographic regions by influencing mosquito breeding sites, which leads to disease transmission.

The direct and indirect costs of malaria in sub-Saharan Africa exceed $2 billion. According to UNICEF, the average cost for each nation in Africa to implement a malaria-control program is estimated to be $300,000 a year, which amounts to about six U.S. cents ($.06) per person for a country of 5 million people. Regrettably, limited resources often prevent such an investment, so malaria lives on.

Prevention and Cure

Malaria prevention needs to be approached from two fronts: protecting against infection and reducing the development of the disease in infected people.

Protective measures against the mosquito have been used for hundreds of years. The inhabitants of swampy regions in ancient Egypt were known to sleep in tower-like structures out of the reach of mosquitoes; others slept under nets. Current measures that protect against infection are still mosquito-focused, like protective clothing, repellents, bed nets, or mosquito control programs. Studies in Ghana, Kenya, and other African nations show that about 30 percent of child deaths could be avoided if children slept under bed nets regularly treated with insecticides.

Mosquito bed net installed in a hut in the village of Kiyi, Kuje, near Abuja, Nigeria.

(© WHO/Pierre Viyot)

Clinic worker in Kiyi Village showing villagers how to soak a bed net in insect repellant.

(© WHO/Pierre Viyot)

The discovery of the insecticide DDT in 1942 and its first use in Italy in 1944 made the ideal of global eradication of malaria seem possible. In the 1950s and 1960s, DDT lowered malaria rates in many parts of the world. However, due to environmental toxicity, it was abandoned. Predictably, malaria rates increased. DDT has been replaced by newer insecticides, which have fewer health risks.

Disease management through early diagnosis and prompt treatment is fundamental to malaria control. Even with the emergence of drug resistance, malaria is largely a curable disease.

Efforts to Combat Malaria

Malaria eradication has been a priority for the World Health Organization (WHO) since its founding in 1948.

The hope of global eradication of malaria was abandoned in 1969. The virus's and the mosquito carrier's ability to develop resistance to drugs and insecticides has contributed to the failure, but social and political factors were, and remain, the biggest barriers. Despite the setbacks, many countries, including Hungary, Bulgaria, Romania, Yugoslavia, Spain, Poland, Italy, Netherlands, and Portugal, eradicated endemic malaria.

A Malaria Vaccine?

The continued existence of malaria, caused by both drug-susceptible and drug-resistant parasites, indicates the need to develop a vaccine to control this disease. Today, there is no generalized vaccine for malaria. This may not be a surprise given the mosquito-host-parasite complexity; however, this complexity provides an opportunity to stop the disease at different stages.

Antigen Alert

Malaria is a treatable disease, and yet it still kills over a million people each year. It is not the medical capability that is missing. In most countries, it is a lack of money and lack of an effective public health system to deliver and administer drugs effectively.

There is optimism that a vaccine is possible, because people living in endemic areas develop immunity to the disease after repeated exposures. Once infected, people rarely develop severe cases of malaria after the age of ten, even though they may carry the parasites in their blood or liver. This immunity is specific for a given strain and wanes rapidly once a person leaves a specific area. Naturally acquired immunity appears to reflect a balance between the parasite and the host's immune system.

An ideal malaria vaccine would prevent infection by stimulating the immune system to destroy all parasites, whether free-swimming in the blood, in the liver, or in red blood cells.

The Least You Need to Know

- Malaria, a curable disease, still kills over a million people a year worldwide.
- People get malaria when they are bitten by certain types of mosquitoes that carry the malaria parasite.
- It is important for those traveling to areas where malaria is endemic to protect themselves.
- New approaches are still needed to combat malaria.

Chapter 10

Sleeping with the Enemy: Sexually Transmitted Diseases

In This Chapter

- ◆ The most common sexually transmitted diseases
- ◆ Symptoms, diagnosis, and treatment
- ◆ How to limit the risk for developing an STD
- ◆ What to do if you think you have an STD

Sexually transmitted diseases (STDs), formerly referred to as "venereal diseases," are among the most common infectious diseases in the world. Amazingly, an estimated 333 million new cases of curable sexually transmitted diseases occur each year among adults. The United States has the highest rate of STDs in the industrialized world, exceeding other nations by 50 to 100 times.

These diseases exert a high emotional toll on afflicted individuals, as well as an economic burden on our healthcare system. More than 20 STDs including chlamydia, gonorrhea, syphilis, genital warts, genital herpes, and viral hepatitis have now been identified as affecting men and women of all backgrounds and economic levels.

Our Current Situation

There are an estimated 15.3 million new cases of STDs in the United States each year, 3 million of which occur in people between the ages of 13 and 19. Nearly two thirds of all STDs occur in people younger than 25! The incidence of STDs is rising partly because people have become sexually active earlier, with more frequent sex partners. Many STDs cause no symptoms, and those that do may be confused with other diseases not transmitted through sexual contact. An infected person, whether he or she has symptoms or not, may transmit an STD to a sex partner. For this reason, periodic screening is recommended for individuals with multiple sex partners.

STDs tend to be more severe for women than for men. In some cases, STDs may spread to the uterus and fallopian tubes, causing pelvic inflammatory disease, a major cause of both infertility and ectopic pregnancy (when the embryo grows in the fallopian tube instead of the uterus). STDs in women are also associated with some cervical cancer. STDs can be passed from a mother to baby before, during, or immediately after birth.

The good news is, however, that when diagnosed and treated early, many STDs can be managed effectively or cured.

> **Disease Diction**
>
> A **sexually transmitted disease** is an infectious disease that is acquired through some type of sexual contact. Once acquired, it can be passed on to other sexual partners.

> **Antigen Alert**
>
> Syphilis, herpes simplex virus, chancroid, and other infections that cause genital or rectal ulcers are associated with HIV infection. Open sores in the genital area are believed to increase the risk for acquiring HIV because they provide an easy entryway for the virus.

Reducing the Chance of Infection

The only completely effective means of protection against STD infection is to abstain from sexual intercourse. For most of the sexually active world, this is an impractical solution.

So for those people engaging in intercourse, the best way to minimize risk is to …

- Have a mutually monogamous sexual relationship with an uninfected partner.
- Use barrier methods of contraception, such as condoms.

- Avoid anal intercourse.
- Delay having sexual relations as long as possible, as young people tend to be more susceptible to infections.
- Keep the total number of sex partners at a minimum.
- Have regular checkups even in the absence of STD symptoms.

If an STD infection is suspected, a doctor should be consulted. Information is readily available from local health departments, as well as from STD and family planning clinics.

Chlamydia: The Most Common STD

Chlamydia trachomatis is the bacterium that causes the most common sexually transmitted disease (STD) in the United States today. It can infect the penis, vagina, cervix, urethra, or eye. The CDC estimates that a staggering four to eight million new cases occur each year. The highest rates of infection are among 15- to 19-year-olds. Symptoms include abnormal discharge (mucus or pus) from the vagina or penis or pain while urinating. Early symptoms may be very mild and usually appear one to three weeks after infection. Often people with chlamydia have few or no symptoms of infection and fail to get treated.

Chlamydia can be transmitted during vaginal, oral, or anal sexual contact with an infected partner. In addition, a pregnant woman may pass the infection to her newborn during delivery, with subsequent neonatal eye infection or pneumonia. Chlamydia can also lead to premature birth or low birth weight.

The bacteria can infect the throat from oral sexual contact with an infected partner. It can also cause an inflamed rectum and inflammation of the lining of the eye ("pink eye"). Complications from chlamydia infections can cause pelvic inflammatory disease (PID), a serious complication that is a major cause of infertility and ectopic pregnancies among women of childbearing age. Each year, approximately 500,000 women in the United States develop PID due to chlamydia infections.

Chlamydia infection is easily confused with gonorrhea because the symptoms of both diseases are similar. Direct culture from the vagina or penis, or from urine, is used to detect the disease. Once diagnosed with chlamydia infection, a person can be effectively treated with antibiotics.

Chlamydia is easy to treat, but both partners must be treated at the same time to prevent reinfection. Antibiotics such as doxycycline, tetracycline, and Zithromax are all effective against chlamydia. Erythromycin is often prescribed for pregnant women and others who cannot take tetracycline.

Potent Fact _____

Hot Numbers:

CDC National STD Hotline
1-800-227-8922 or
1-800-342-2437
En Espanol 1-800-344-7432
TTY 1-800-243-7889

National Herpes Hotline
919-361-8488

National HPV and Cervical Cancer Hotline
919-361-4848
www.ashastd.org/hpvccrc/

CDC National Prevention Information Network
P.O. Box 6003 Rockville MD 20849-6003
1-800-458-5231 or
1-888-282-7681
Fax 1-800-243-7012
www.cdcnpin.org/
info@cdcnpin.org

Symptoms for women include:

- Bleeding between menstrual periods
- Vaginal bleeding after sex
- Abdominal pain
- Painful intercourse
- A low-grade fever
- A painful sensation during urination
- The urge to urinate more than usual
- Abnormal vaginal discharge
- A foul-smelling yellowish discharge from the cervix
- Eye infections

Symptoms for men include:

- Pus or milky discharge from the penis
- Pain or burning sensation when urinating
- Fever
- Genital Swelling
- Watery, white, or yellow drip from the penis
- Extreme pain in the scrotum
- Eye infections

Chlamydia can spread from the urethra to the testicles causing a condition known as "epididymitis." Reiter's syndrome, a common type of arthritis due to inflammation of the joints, has also been linked to chlamydia infections in young men.

Genital Herpes: Lifelong Infection

Herpes simplex virus (HSV), better known as genital herpes, is a contagious viral infection estimated to infect 45 million Americans, with as many as 500,000 new cases occurring each year. Infections frequently go unrecognized by patients and/or clinicians. Two types of virus, HSV1 and HSV2, cause genital herpes. Both types produce sores in and around the vagina, penis, anal opening, and on the buttocks or thighs. Sores may also appear on other areas whenever broken skin comes into contact with HSV.

The virus invades nerves cells and can reside there for life, causing periodic symptoms. Genital herpes infection is acquired by sexual contact with a partner having an outbreak of herpes sores in the genital area. Oral herpes can be transmitted to the genital area of a partner during oral sex. Some herpes infections may make people more likely to get an HIV infection if exposed to the virus. Reliable tests for HSV antibodies are now readily available. In addition, PCR tests can be used to detect herpes infection.

There is no cure for herpes. However, there are a number of drugs that are effective in treating the herpes virus. Acyclovir, an antiviral drug, is the "gold standard" of therapy. These drugs reduce symptoms and help to speed healing. They also lessen the chances of outbreaks. There is no vaccine for genital herpes, although recent trials of vaccines reduced the risk of infection by 75 percent. The vaccine was not effective in men, however, making it the first time a vaccine worked in one sex and not in the other. Unfortunately, herpes can be spread even if the infection is inactive.

Potent Fact

Genital herpes is not readily spread by contact with a toilet seat or in a hot tub.

Genital Warts and HPV

Genital warts are growths or bumps that appear on the vulva, vagina, anus, cervix, penis, scrotum, or thigh. They may be raised or flat, single or multiple, small or large, or clustered to form cauliflower-like shapes. The warts are caused by the human papillomavirus (HPV), which also causes cervical cancer and other genital cancers.

HPV is one of the most common causes of sexually transmitted disease in the world. Nearly 24 million Americans are infected with HPV. Some cause common skin warts, while about one third of the HPV types are transmitted by sexual contact and reside in genital tissue without causing warts. HPV infection often shows no symptoms—it is estimated that almost half of the women infected with HPV have no obvious symptoms.

Like other STDs, individuals with HPV infection are largely unaware of the potential risk they pose for transmission to others. More than 80 HPV types have been identified, many of which are known to cause genital warts. Some types also have been associated with cancer.

How HPV Is Spread

Genital warts are transmitted by direct, skin-to-skin contact during vaginal, anal, or oral sex with an infected partner. A common theory is that the virus enters the genital tissue through micro-abrasions caused by sexual activity.

HPV and Cancer

Incredibly, HPV can be detected in 93 percent of all cervical cancers, yet the virus alone is not sufficient to induce cervical cancer. Other factors include oral contraceptive use, poor nutrition, a weakened immune system, pregnancy, and smoking. Although most HPV infections do not progress to cancer, it is particularly important for women who have had evidence of HPV infection or genital warts to have regular *Pap smears*.

Disease Diction

In a **Pap smear,** cells are scraped off the cervix and sent to a laboratory to be examined under a microscope. The cells are examined for the presence of the HPV virus, which can be an indicator of cervical cancer.

Checking Up on HPV

Visible warts can be diagnosed by directly observing the genital area. Infections without visible warts can be diagnosed by using a mini-magnifying scope or through blood tests. Latent infection, when the disease isn't actively producing warts, can be diagnosed only by detecting HPV by PCR testing. (See Chapter 5 for more on PCR testing.) If a person has genital warts that reappear quickly after treatment, or warts that have pigment and are larger than one centimeter in diameter, their doctor should check the tissue by doing a biopsy to test for cancer.

Approximately 25 percent of patients with genital warts are infected with another STD, and patients should be screened for chlamydia, gonorrhea, syphilis, and HIV. The best way to diagnose HPV infection in women is via a Pap smear test, as part of a regular health check-up.

There Isn't a Cure ...

Scientists have not found a cure for HPV—once someone contracts the virus, it is with them for life. The goal of treatment is to remove warts and to reduce other overt symptoms of the infection. The genital warts themselves can be removed by surgically freezing them, burning them with acid or lasers, or applying prescription creams, but removing the warts doesn't eradicate the virus.

Pregnancy/Birth

Genital warts tend to grow rapidly during pregnancy, probably because of the woman's suppressed immune system. Genital warts may cause a number of problems during pregnancy. If the warts are in the vagina, they can make the vagina less elastic and cause obstruction during delivery. Warts in the throat is probably the most life-threatening complication to a fetus exposed to HPV during vaginal delivery.

Avoiding Infection

The only way to prevent HPV infection is to avoid direct contact with the virus. If warts are visible in the genital area, sexual contact should be avoided until the warts are treated. Use of barrier protection (condoms) during sexual intercourse may provide some protection, but is not a guarantee.

Unprotected sex is the most common contributing factor to infection with HPV and greatly increases chances for developing of cervical dysplasia, which are precancerous changes of the cervix. Other sexual behaviors, such as multiple sex partners and sexual intercourse at an early age, increase the risk for cervical dysplasia, too. HPV infection can be spread even if warts aren't present.

Syphilis: A Sexual Scourge with a Long History

Syphilis is a sexually transmitted disease that begins with genital sores, progresses to a general rash, and then to disfiguring abscesses and scabs all over the body. In its late stages, untreated syphilis can cause heart abnormalities, mental disorders, blindness, other neurological problems, and death. It appeared prominently in Europe at the end of the fourteenth century, and by 1500 syphilis had spread to much of the continent. The explorer Vasco da Gama carried it to Calcutta in 1498, and by 1520 syphilis had reached Africa and China. It was considered the sexual scourge of the sixteenth century.

For centuries, syphilis remained a major component of the infectious disease landscape throughout Europe, Asia, and Africa. The United States has been no stranger to the disease, which most likely arrived with fifteenth- and sixteenth-century explorers. The rate of syphilis peaked in the U.S. in 1947 at 106,000 cases, but was dramatically reduced following the widespread introduction of antibiotics.

By 1996 in the United States only 11,387 cases of syphilis in its infectious stages were reported to the CDC. African Americans are 34 times more likely to be reported with syphilis than whites. In October 1999, the CDC launched a national campaign to eliminate syphilis in the United States by 2005. Since then, a steady decline in the number of people infected with syphilis reflects the positive efforts of this program.

> **Infectious Knowledge**
>
> Al Capone, one of the most notorious gangsters of all time, contracted syphilis around 1927. After being imprisoned on Alcatraz Island, Capone's syphilis was far advanced and he was reduced to a babbling idiot despite treatment from prison doctors.

The Cause

Treponema pallidum is the bacterium that causes syphilis. Syphilis can move throughout the body, damaging many organs over time. After initial penetration, the bacteria enter the lymph capillaries, where they are transported to the nearest lymph gland. There, they multiply and are released into the bloodstream, where they invade every part of the body.

Infectious Knowledge

Microbiologists Schmudinn and Hoffman in 1905 discovered and isolated the bacterium that causes syphilis. In 1906 German bacteriologist August von Wassermann, working in conjunction with Albert Neisser, discovered the Wassermann reaction, a blood-serum test that could determine if a person had syphilis. German scientist Paul Ehrlich, in 1908, began his research to find a better drug to fight the disease by testing hundreds of different arsenic compounds on syphilitic rats. One compound was found that effectively destroyed syphilis without destroying the rat. He called it Salvarsan, which in English means "I save."

How Syphilis Is Spread (Watch Out for Sores!)

In acquired syphilis, the bacterium enters the body through skin or mucous membranes, usually during sexual contact. It can be spread from the sores of an infected person to the mucous membranes of the genital area, the mouth, or the anus of a sexual partner. The bacteria are very fragile, and the infection is not spread by contact with objects such as toilet seats or towels.

Congenital syphilis occurs when a pregnant woman with syphilis passes the bacteria through the placenta to her unborn child. People at risk for syphilis, like other STDs, are those who have had multiple sex partners and engage in high-risk sexual practices. An infected person who does not get treatment may infect others during the first two stages of the disease (see the next section) when sores are present.

Syphilitic Symptoms

Syphilis is characterized by four prominent stages: primary, secondary, latent, and tertiary (also called the late stage). The primary stage is marked by the appearance of a sore, called a chancre, which usually occurs 10 to 21 days following exposure but may take up to three months to appear. The chancre is generally found on the penis, vagina, or rectum but may also develop on the cervix, tongue, lips, or other parts of the body. It is usually painless and disappears within a few weeks regardless of whether or not it is treated.

Secondary syphilis is distinguished by a skin rash that appears up to 10 weeks after the chancre heals. The rash, resembling measles or chickenpox, may cover the whole body or appear only in a few areas, such as on the palms of the hands or soles of the feet. Fever, indigestion, or headaches may accompany the rash. The rash usually heals within several weeks or months. In some cases, ulcers may appear in the mouth. Scalp hair may drop out in patches, creating a "moth-eaten" appearance. Other symptoms include mild fever, fatigue, headache, sore throat, and swollen lymph glands. Pain in bones and joints and heart palpitations may develop. However, the symptoms, which generally tend to be mild, disappear without treatment but may reoccur over the next two years.

Next, the disease goes into a latent or hidden stage when it is difficult to detect the bacteria even with tests. If left untreated during the latent stage, 50 to 70 percent of people who carry the disease suffer no further consequences. However, approximately one third of those infected develop complications of late, or tertiary, syphilis in which the bacteria damage the heart, eyes, brain, nervous system, bones, joints, or almost any other part of the body. This stage can last for years, or even decades. Late syphilis is the feared stage, because it can result in mental illness, blindness, other neurological problems, heart disease, and death.

> **CAUTION**
>
> **Antigen Alert**
>
> Syphilis can invade the nervous system, causing neurosyphilis, which may result in headaches, stiff neck, and fever from inflammation of the lining of the brain. Neurosyphilis may cause paralysis and insanity, as well as a degeneration of the spinal cord, causing a stumbling, foot-stamping gait.

Congenital Syphilis: Infection During Pregnancy

A pregnant woman with active untreated syphilis will pass the infection to her unborn child. Up to 70 percent of such pregnancies will result in a syphilitic infant. Most babies develop symptoms such as skin sores, rashes, fever, weakened crying, swollen liver and spleen, jaundice (yellowish skin), anemia, and various deformities between two weeks and three months after birth. Sometimes the symptoms of syphilis don't appear for many years, and older children may develop symptoms of late congenital syphilis, including damage to their bones, teeth, sight, hearing, and brain.

Diagnosing "the Great Imitator"

Syphilis has sometimes been called "the great imitator" because its early symptoms resemble those of many other diseases. People who have more than one sex partner should consult a physician about any suspicious rash or sore in the genital area. Individuals who have been treated for another STD, such as gonorrhea, should be tested for syphilis.

Syphilis can be diagnosed by evaluating a combination of clinical signs and symptoms, direct microscopic observation in tissue or blood, and antibody-based blood tests. The principal screening tests for syphilis are the VDRL (Venereal Disease Research Laboratory) and RPR (rapid plasma regain) tests, which detect a rise in antibody following infection. The later the stage of infection, the greater the ability of the tests to detect antibodies.

While blood tests can provide evidence of infection, they may give false negative results for up to three months after infection. In addition, blood tests for syphilis can sometimes be positive even though a person isn't infected with the disease. Interpretation of blood tests for syphilis can be difficult, and repeated tests are sometimes necessary to confirm the diagnosis.

With Proper Treatment, There Is a Cure

At one time, mercury was given to people suffering from syphilis, but it probably poisoned more people than it helped. Today, syphilis is treated with penicillin. Other antibiotics, such as tetracycline, can be used for patients allergic to penicillin. It is important that people being treated for syphilis have periodic blood tests to ensure that they have been cured.

Infectious Knowledge

In 1932, the United States Public Health Service, in cooperation with the Tuskegee Institute, initiated a study in Macon County, Alabama, to determine the effects of untreated syphilis. From 1932 through 1972, 399 low-income African American men with latent syphilis went untreated for the disease after being offered "so-called" free medical care. The men, the most educated of whom completed seventh grade, were told they were being treated for an ongoing condition of "bad blood." They were never told that they were part of a study, nor were they informed that they had syphilis. Government doctors failed to offer standard treatments, nor did they offer penicillin once it became the standard method of curing the disease. The study was stopped in 1970 only after its existence was leaked to the public. By that time, at least 28, and perhaps as many as 100, men had died as a direct result of complications caused by syphilis.

The surviving men received treatment only after the experiment became public. In December of 1974, the government agreed to pay approximately $10 million in an out-of-court settlement: $37,500 per participant. The lessons learned from this "holocaust era" experiment formed the basis for current-day human subjects research guidelines. On May 16, 1997, the surviving participants of the Tuskegee Syphilis Study gathered at the White House and witnessed President Clinton's long overdue apology on behalf of the United States government.

Individuals with syphilis that has invaded the nervous system may need to be retested for up to two years after treatment. Fortunately, proper treatment will cure the disease in all stages.

The Best Protection

Not having sex is the best protection against acquiring syphilis and other STDs. Open sores associated with syphilis may be visible and are infectious during the active stages of the disease. Any contact with these infectious sores must be avoided to prevent the spread of the disease. Latex condoms can reduce the risk of syphilis; however, lesions may occur in areas that cannot be covered or protected by a condom. Testing and treatment early in pregnancy is the best way to prevent syphilis in infants.

There is no vaccine for syphilis, no acquired immunity to syphilis, and past infection provides no protection! A person can be infected with syphilis, treated, and then become reinfected, once, twice, or a hundred times.

Yet Another Nasty STD: Gonorrhea

Gonorrhea is a curable sexually transmitted disease caused by the bacterium *Neisseria gonorrhoeae*. It infects approximately 750,000 people in the United States each year, although another 750,000 unreported cases are also believed to occur each year. According to the CDC, the rate of reported gonorrhea infections was 132.2 per 100,000 persons in 1999.

Any sexually active person can be infected with gonorrhea, yet nearly 75 percent of all reported gonorrhea is found in individuals 15 to 29 years old.

The most common symptoms of gonorrhea are a discharge from the vagina or penis and painful or difficult urination. It can infect the genital tract, the mouth, and the rectum. In women, the cervix and uterus can be the first place of infection, with the disease later spreading to the uterus and fallopian tubes, resulting in pelvic inflammatory disease (PID). PID affects more than one million women in the U.S. each year and can cause ectopic pregnancies and infertility in as many as 10 percent of infected women.

> **CAUTION**
>
> **Antigen Alert**
>
> The highest rates of infection are found in 15- to 19-year-old women and 20- to 24-year-old men, with a disproportionate number of cases (77 percent of the total number) reported among African Americans.

Burning, Pain, and Swelling

In men, symptoms include a burning sensation when urinating and a yellowish white discharge from the penis. Some men experience painful or swollen testicles.

In women, the early symptoms of gonorrhea are often mild, with few, if any, symptoms of infection. Like other STDs, gonorrhea can be mistaken for a bladder or vaginal infection. The initial symptoms and signs in women include a painful or burning sensation when urinating and a vaginal discharge that is yellow or occasionally bloody. If gonorrhea is not treated, the bacteria can spread to the bloodstream and infect the joints, heart valves, or the brain. Untreated gonorrhea in women can develop into pelvic inflammatory disease (PID). Rectal infection in men and women includes discharge, anal itching, soreness, bleeding, and sometimes painful bowel movements.

Doctors test for gonorrhea by sending a sample consisting of a vaginal or penile discharge, or fluid from the infected mucous membrane (cervix, urethra, rectum, or throat) to a lab for analysis. Infections present in the male or female genital tract can be diagnosed from a urine specimen. The bacteria can also be identified using a microscope. More advanced testing involving the detection of bacterial genes or nucleic acid (DNA) in urine, and growing the bacteria in laboratory cultures are the most accurate.

Treatment and Drug-Resistance

Gonorrhea can be effectively treated with antibiotics. Historically, penicillin has been used to treat gonorrhea, but ampicillin and amoxicillin are also prescribed. Unfortunately, strains of penicillin-resistant gonorrhea are increasing, and newer antibiotics or combinations of drugs must be used to treat these resistant strains.

Gonorrhea and chlamydia infection often infect people at the same time; therefore, some physicians prefer a combination of antibiotics such as ceftriaxone and doxycycline or azithromycin, which will treat both diseases.

Precautions and Prevention

People diagnosed with gonorrhea need to tell all of their sexual partners, so they can get tested and then treated if infected, whether or not they have symptoms of infection. If a woman has gonorrhea when she gives birth, the infection can be passed to the newborn and cause eye damage.

People with gonorrhea can also more easily contract HIV. The best prevention is to practice sexual abstinence. Sexually active individuals should engage in safe sex practices including the use of barrier protection methods, like condoms. However, condoms do not provide complete protection from all STDs. Sores and lesions of other STDs on infected men and women may be present in areas not covered by the condom, resulting in transmission of infection to another person. Limit the number of sex partners, and do not go back and forth between partners. If you think you are infected, avoid sexual contact and see a health-care provider immediately.

Sexually transmitted diseases are clearly a major problem, and they are often under-diagnosed. It is important to try to practice safe sex and to see a doctor if any symptoms appear. Although many of these diseases aren't life-threatening, if left untreated they can cause fertility problems and have other unpleasant long-term effects.

> **Infectious Knowledge**
>
> There is no effective vaccine against gonorrhea largely because it is a complex organism that infects only humans, and it has a remarkable ability to protect itself by changing or mutating.

The Least You Need to Know

◆ Sexually transmitted diseases are very common and affect a large percentage of the population at some point in their lives.

◆ Most sexually transmitted diseases are treatable with antibiotics or antiviral drugs.

◆ Limiting the number of sexual partners and using condoms can help lower the risk for getting an STD.

An Alphabet Soup of Infections: Viral Hepatitis

In This Chapter

- ◆ The six varieties of viral hepatitis
- ◆ Acute vs. chronic infection
- ◆ Diagnosis and treatment
- ◆ The long-term outlook

Hepatitis is a disease of the liver that causes inflammation and swelling, potentially resulting in permanent damage. It can be quite painful and its effects can make its victims feel weak. Hepatitis infections can be acute or chronic, and people can die from them.

The disease can be caused by many different factors, including infectious organisms, chemical toxins, poisons, drugs, and alcohol. Unlike nonviral hepatitis, which can be caused by any number of noninfectious means, viral hepatitis is caused by one of six different viruses: hepatitis A, hepatitis B, hepatitis C, hepatitis D, hepatitis E, or hepatitis G.

Viral hepatitis may spread through contaminated food or water, or it may be blood-borne. The worldwide incidence of viral hepatitis is enormous, with more than 300 million carriers of hepatitis B and 170 million cases of hepatitis C.

Hepatitis A: Thriving in Unsanitary Conditions

Hepatitis A is transmitted by human consumption of fecal-contaminated drinking water or food. Like cholera, the risk of contracting hepatitis A depends on the hygiene and sanitary conditions in a given area. Developing countries are at high risk, although 180,000 people in the United States are infected each year by hepatitis A. About 100 people in the United States die from hepatitis A-related complications each year.

Antigen Alert

When traveling to an area known to have hepatitis, always emphasize personal hygiene. And don't forget the basics of food preparation and consumption: boil it, broil it, peel it, or forget it!

A high concentration of the virus is found in fecal matter, and the virus can survive on the hands or other surfaces for up to four hours at room temperature. Eating utensils are a frequent source of infection, as are contaminated shellfish. Hepatitis A can also be spread through intravenous drug use and sexual contact.

Symptoms and Treatment

A common characteristic of hepatitis is that the infected person may not have any symptoms. When symptoms do occur, usually within the first four weeks of infection, they may be flu-like, with fatigue, body ache, nausea, vomiting, pain, and tenderness in the liver area, dark urine, or light colored stools and fever.

Other indications are jaundice in adults, where the skin and eye color take on a yellow hue, and liver test results that indicate a higher level of activity of key enzymes than normal. There is no specific treatment for hepatitis A. Supportive care is recommended and is guided by symptoms, which often last from about four weeks to a few months. Symptoms may return in 20 percent of people who get the disease and continue on and off for up to 15 months. However, the infection will resolve by itself, with no serious after-effects. Once recovered, an individual is then immune to reinfection. Only about one percent of all hepatitis A infections cause a severe enough infection that damages the liver to an extent that a transplant is required.

Testing for Hepatitis A

There are currently two blood tests available to detect hepatitis A. Antibodies may be detected for up to six months following the onset of symptoms, but they tend to disappear after time.

Getting Vaccinated

There are two approved vaccines available in the United States for protection against hepatitis A. They provide long-term protection and are licensed for use in children two years of age and older. Two doses are needed, 6 to 12 months apart, to ensure long-term protection. International travelers should get the first dose at least four weeks prior to their departure.

Immunoglobulin (plasma containing different classes of antibodies made from people who are immune to the disease) is recommended for short-term protection, but it must be given within two weeks of exposure for maximum protection. People born and raised in developing countries where hepatitis A is endemic have usually been infected in childhood with a mild case, and are generally immune to the disease.

Potent Fact

Hepatitis A is the most common vaccine-preventable disease in international travelers. It is 1,000 times more common than cholera and 100 times more common than typhoid among international travelers.

Hepatitis E: The Hepatitis A Copycat

Hepatitis E disease has symptoms much like hepatitis A. It is an acute, short-duration disease spread widely in many tropical and underdeveloped countries, usually through contaminated drinking water. Hepatitis E affects young adults rather than children, and causes a high death rate, particularly in pregnant women. Major waterborne epidemics have occurred in Asia and North and East Africa, but there have been no known outbreaks in the United States besides some sporadic cases in Los Angeles in 1987.

Like hepatitis A, good sanitation and personal hygiene are the best preventive measures. The incubation period for hepatitis E varies from two to nine weeks, with the disease itself usually lasting about two weeks. Although infections are generally mild for young adults, the fatality rate in pregnant women approaches 20 percent. Protection from the disease in endemic areas lies mainly in prevention, as a vaccine for hepatitis E is in the experimental stage. Hepatitis E can be diagnosed based on symptoms and the elevated presence of liver enzymes.

Potent Fact

Having access to clean drinking water, washing hands before eating, and proper disposal of sewage are still the best ways to decrease the incidence of this disease.

Hepatitis B: 300 Million Carriers and Growing

Hepatitis B is highly prevalent in the United States, with approximately 1.2 million chronic carriers. Hepatitis B is the ninth leading cause of death worldwide, and there are more than 300 million carriers. Hepatitis B may develop into a chronic disease in up to 10 percent of newly-infected people each year. If left untreated, the risk of developing *cirrhosis* (scarring) and liver cancer becomes higher.

> **Disease Diction**
>
> **Cirrhosis** of the liver is the scarring and death of liver tissue. Cirrhosis occurs as a result of alcohol, malnutrition, or hepatitis. When it happens, the liver doesn't function normally, fluid can build up in the abdomen, blood pressure can increase, and sometimes brain disorders can occur.

How You Get It

Hepatitis B is highly infectious and is transmitted through infected blood and other body fluids (seminal fluid, vaginal secretions, breast milk, tears, saliva, and open sores). Like many other sexually transmitted diseases, hepatitis B is spread through unsafe sexual practices. Health-care workers, prison personnel and inmates, and intravenous drug users are at particularly high risk.

The blood supply in the United States has been screened for hepatitis B for many years, and transfusion-related illness is extremely rare; however, recipients of blood or blood products before 1975, when screening started, are at risk. Hepatitis B is transmitted from mother to infant in the last few months of pregnancy, which is a major mode of transmission in regions where the disease is endemic.

What It Feels Like

Most people who get hepatitis B have no recognizable signs or symptoms. When they do appear, symptoms include flu-like illness, loss of appetite, nausea and vomiting, fever, weakness, and mild abdominal pain. Dark urine and jaundice may also be observed. Most adults (90 to 95 percent) recover within six months, while 5 to 10 percent develop chronic hepatitis or become carriers. About 50 percent of infected young children will become chronically infected.

Chronic infection can result in cirrhosis, liver failure, and death in severe cases. Liver cancer is also associated with hepatitis B. Hepatitis B carriers are potentially infectious even though they have no symptoms.

Chronic infection comes in two forms. In the first, the virus multiplies rapidly and is easily detected. In the second, there are low rates of viral replication and it is difficult to detect. Patients with the first form of the disease generally have a worse prognosis and a greater chance of developing cirrhosis and cancer.

Detecting B

Both blood and molecular tests are useful in the diagnosis of viral hepatitis. They may detect early infections before other signs of disease appear, and can distinguish between acute and chronic infections.

Nearly all infected individuals will have detectable hepatitis B antigens. Acute infection is diagnosed by the presence of both antigens and antibodies; the antibody develops in the early stages of infection, at the time symptoms appear (the same antibodies are also produced by vaccination). Most people with acute infection retain some antibodies, but in some cases, the antibody response wanes over time. This makes it difficult to diagnose reinfection. Diagnosis of hepatitis B should be confirmed by a liver biopsy.

Hepatitis B-Associated Cancer

Hepatocellular carcinoma (HCC), a form of liver cancer, is the most common malignant tumor found in males worldwide, with one million new cases each year, mostly in Southeast Asia and Sub-Saharan Africa. The incidence of this cancer is closely associated with hepatitis B infection in endemic regions. Chronic alcohol consumption and cirrhosis, along with chronic viral infection, increase the development of cancer. Cancer screening should include an ultrasound examination.

Treating B

Although there is no treatment for acute hepatitis B, there are two approved treatments for chronic hepatitis B, the first being *alpha interferon*. The exact way that alpha interferon works is unknown, but it is thought to restrict viral replication and attachment and boost T cells. Unfortunately, people with a weakened immune system do not respond well to therapy. Alpha interferon therapy often results in a number of side effects, including flu-like symptoms, fatigue, headache, nausea and vomiting, loss of appetite, depression, and hair thinning. Because interferon may depress the bone marrow, blood tests are needed to monitor white blood cells and platelets.

The second treatment, a drug named Lamivudine, stops viral replication. Drug resistance can occur, but usually after 9 to 12 months of exposure and at

Disease Diction

Alpha interferon is a naturally occurring protein secreted by cells in response to viral infections. It enhances the production of certain chemicals in the body that help to boost the immune response. It also inhibits viral replication in infected cells.

Potent Fact

If an accidental exposure to hepatitis B occurs in a person who is not immune, an effective response is to treat that person with hepatitis B immunoglobulin antibodies from people who are already immune to infection with hepatitis B.

a rate of only 10 to 15 percent per year. A once-a-day tablet, Lamivudine is less expensive and has fewer severe side effects than alpha interferon. The most common side effects are fatigue, headache, nausea, and abdominal pain. Neither therapy is considered great because of the side effects, but they are currently the only two available.

Antigen Alert _____

The following series of events is from a CDC Case Report:

♦ July 1999: A hospital stops giving newborns the first dose of hepatitis B vaccine because of health concerns about a preservative (thimerosal) it contains.

♦ September 1999: Hepatitis B vaccine without thimerosal as a preservative becomes available, but the hospital elects not to resume routine neonatal hepatitis B immunization.

An infant born at the hospital receives neither hepatitis B vaccine nor hepatitis B immunoglobulin. The infant develops hepatitis B at 3 months of age and dies less than 2 weeks after the onset of symptoms.

This tragedy is preventable and should never happen again.

There Is a Vaccination, But It's Expensive

Effective vaccines are available for the prevention of hepatitis B infection. Currently two forms of vaccines are available, both of which are injectable and expensive; two or three injections over a 6- to 12-month period are required to provide full protection. Once complete, the vaccines provide protection against hepatitis B for at least 15 years.

All individuals at risk for infection should be vaccinated, as well as all children and adolescents, because most cases occur in sexually active young adults. Chronic infection remains a major problem worldwide, despite some declines due to the hepatitis B vaccine. Patients with chronic hepatitis B should also be vaccinated against hepatitis A.

Infectious Knowledge

Infection with hepatitis D occurs only in patients already infected with hepatitis B. Hepatitis D is spread mainly by contaminated needles and blood. The simultaneous infection with hepatitis B and hepatitis D produces more severe illness and higher rates of long-term liver failure than hepatitis B alone. The disease usually goes away on its own; and due to its codependence on hepatitis B, hepatitis D is effectively prevented via the hepatitis B vaccine.

Hepatitis C: The Silent Killer

The hepatitis C virus causes both acute and chronic liver disease. Prior to its identification in 1989, large numbers of hepatitis victims were found to be negative for both hepatitis A and B. The unknown disease was known as non-A, non-B hepatitis before being named hepatitis C.

Researchers quickly discovered that hepatitis C accounted for large numbers of hepatitis cases, and it has been identified as a modern pandemic. An estimated three percent of the world's population is chronically infected with hepatitis C, including four million people in the United States, making it one of the greatest public health threats of this century. Unlike other types of hepatitis, more than 80 percent of hepatitis C infections become chronic and lead to liver disease. Hepatitis C, in combination with hepatitis B, now accounts for 75 percent of all cases of liver disease around the world. Liver failure due to hepatitis C is the leading cause of liver transplants in the United States. Chronic hepatitis C is a major cause of cirrhosis and liver cancer, which most often lead to liver transplantation.

The virus has six subgroups, some of which are distributed worldwide, while others are found in more restricted areas. Certain racial, ethnic, and income groups are at higher risk of infection, in part due to unidentified modes of transmission. In the United States, African Americans have the highest incidence rates, followed by Native Americans, Hispanics, and whites. Similarly, low-income groups seem to have the highest risk of infection.

What It Feels Like

Most of the time, both acute and chronic hepatitis C have no symptoms. However, chronic hepatitis C is a slowly progressive disease and results in severe disease in 20 to 30 percent of infected people.

The symptoms of hepatitis C are difficult to recognize because they tend to be mild during the early stage of infection. The most common symptom is fatigue, but it may take years to become manifest. Other symptoms include flu-like mild fever, muscle and joint aches, nausea, vomiting, loss of appetite, vague abdominal pain, and sometimes diarrhea. A small number of individuals have dark urine, light-colored stool, and jaundice. Itching of the skin and weight loss (5 to 10 pounds) occur occasionally. Disorders of the thyroid, intestine, eyes, joints, blood, spleen, kidneys, and skin may occur in about 20 percent of patients.

How You Get It

Hepatitis C is transmitted only by blood. Many individuals contracted hepatitis C through blood transfusions prior to the 1990s, when screening for the disease started. Since then, screening has nearly eliminated this route of transmission. Intravenous drug use and high-risk sexual activity are the most frequently identified risk factors associated with hepatitis C infection.

Hepatitis C infection is common in sexually promiscuous individuals, but not generally in monogamous couples. Unlike hepatitis B, C is not spread readily from mother to child during birth. Certain routine activities pose a risk for infection, including manicures, and the sharing of toothbrushes and blade razors. Whether a person progresses to chronic liver disease or not, the infected individual carries the virus for life. This means that they also remain contagious for a lifetime and are able to transmit the virus to others. The long progression of the illness indicates that individuals can carry hepatitis C for decades.

> **Potent Fact**
>
> A very small percentage of patients may recover from acute hepatitis C, but tests will still show the presence of antibodies for the virus in their blood even after they are cured.

Diagnosing It

Hepatitis C infection is rarely diagnosed in its early stages. It is often not recognized until its chronic stages, when it has caused severe liver disease. Hepatitis C infection is often referred to as the "the Silent Epidemic," because a typical cycle of the disease from infection to symptomatic liver disease may take as long as 20 years. The diagnosis of chronic hepatitis C disease is made by blood testing and liver biopsy. In addition, several molecular tests have been developed to detect the virus directly. In general, elevated liver enzymes and a positive antibody test means that an individual has chronic hepatitis C.

Treating It

There are a number of drug treatments becoming available for hepatitis C. The treatment of patients with chronic infection who have not been treated previously generally consists of interferon alpha and ribavirin.

Liver transplantation may be life-saving in end-stage liver disease, but this treatment option is limited by a shortage of liver donors. In addition, reinfection is almost universal in hepatitis C-positive patients undergoing transplantation. Unfortunately, the treatments available for hepatitis C are painful and sometimes worse than the symptoms of the disease itself. The infection lasts a long time, so often people whose treatment is well managed live a long time.

A vaccine against hepatitis C may not be available for many years to come.

Overtaking AIDS

Without prompt intervention to treat infected populations and prevent the spread of disease, the death rate from hepatitis C will soon surpass that from AIDS—and it can only get worse. Hepatitis C can mutate frequently, so different genetic variations can live within the host. The mutated forms are sufficiently different that the immune system does not recognize them. As a result, the development of antibodies against hepatitis C does not produce immunity against the disease like it does with most other viruses.

Antigen Alert

Your mother was right. Tattooing, body piercing, acupuncture, and ear piercing can be dangerous. They contribute to the spread of viruses like hepatitis C.

Living with Hepatitis C

Living with hepatitis C can be difficult, as constant fatigue is a common problem. For many people, frequent, short naps prevent extreme fatigue. It is also helpful to limit stressful activities, although routine activities may appear overwhelming. It is important to avoid stressing a damaged liver by consuming alcohol or ingesting other potentially dangerous substances, including toxins, certain metals like copper, and some over-the-counter drugs such as aspirin.

Hepatitis G: The New Virus on the Block

Hepatitis G is transmitted by blood-borne routes and was just discovered in 1995. Risk groups include intravenous drug users, hemodialysis patients, and transfusion recipients. Acute infection is diagnosed by antibody tests or by molecular techniques like PCR that detect the presence of RNA from the virus.

People can remain infected for many years. The first major study of virus has reported that those infected by means other than blood transfusions did not develop chronic liver disease. Because it was only identified recently, it isn't clear at this time how widespread hepatitis G is and what its precise effects are on infected patients.

Hepatitis H, I, and J?

There is increasing evidence that other hepatitis viruses exist. This is because known viruses do not explain all cases of hepatitis. What this means for future treatment of hepatitis is unknown, however.

The Least You Need to Know

◆ Viral hepatitis comes in a variety of forms. Each form has different symptoms and requires a different treatment.

◆ Vaccines are available for some forms of hepatitis. Travelers to certain parts of the world should be vaccinated two to four weeks before they leave.

◆ Some forms of hepatitis can lead to chronic liver disease and even the need for a liver transplant.

◆ It is important to get tested if you think you have any of the symptoms described.

Part  Hot Viruses and Other Killer Bugs

Don't mess with Mother Nature, particularly if she takes the form of the infectious diseases discussed in this section.

Although so-called "hot viruses" like Ebola and Marburg might be the scene-stealers, more people are actually killed by seemingly mundane diseases such as fungal infections, cholera, dysentery, and yellow fever. Children and people who live in tropical areas are often prey to particularly nasty breeds of diseases, which you'll learn about in this part, too. Finally, join us as we remove the mask from some infectious diseases in disguise—diseases that until recently we didn't know were caused by infectious agents.

Chapter 12

Not-So-Fun Fungal Infections

In This Chapter

- ◆ The nature of fungi
- ◆ Controlling and treating fungal infections
- ◆ Opportunistic fungal infections and how they spread
- ◆ Other infections caused by fungi

The incidence of fungal infections has increased at an alarming rate in the past two decades. Most of this increase is due to opportunistic fungal infections related to the growing population of people with weakened immune systems due to HIV, cancer, and other diseases; and to modern medical practices such as the use of intensive chemotherapy and drugs that suppress the immune system. However, fungal infections like vaginal yeast infections and athlete's foot are common in healthy people, too.

In this chapter, we will cover a variety of fungal infections affecting both those with weakened immune systems and healthy people.

The Fungus Among Us

Fungi are everywhere—as moulds, they grow in homes and on foods; as yeasts, they are found in foods and in our bodies. Even mushrooms are classified as fungi—although most of them are quite harmless. Whatever form fungi take, they survive by breaking down organic matter. Only 180 of the 250,000 known species of fungi can cause disease in people.

When a Good Thing Goes Bad

Many different kinds of fungi live inside the human body in peaceful equilibrium with it. When the body's immune system is weakened due to an illness such as AIDS or treatment such as chemotherapy, the balance can be disturbed, allowing fungi to cause disease. Such diseases are called *opportunistic infections*.

Disease Diction

An **opportunistic infection** is caused by an organism that normally lives in our body or our environment and causes no damage, but takes advantage of the opportunity a weakened immune system gives it to cause disease.

Antigen Alert

It is often difficult to diagnose systemic fungal infections. Many times they are confirmed only at autopsy. Because so many systemic fungal infections go undetected and untreated, antifungal therapy has mixed success.

People most often come into contact with fungi in the organisms' natural habitats. Because many fungi live in the ground, gardeners are often at risk for fungal infections. The organisms can enter the body through bare feet, hands, or other exposed areas. Different kinds of fungi live in different geographic areas. For example, cocci is a disease of the Southwestern United States and is found in people who live in or visit this area.

Types of Infection

There are a number of different types of infections caused by fungi, including …

- **Superficial infections** These fungal infections affect the skin or mucous membranes. Superficial fungal infections (e.g., yeast vaginitis, oral thrush, and athletes foot) affect millions of people worldwide. Although rarely life-threatening, they can have debilitating effects on a person's quality of life and may in some cases spread to other people or become invasive (systemic). Most superficial fungal infections are easily diagnosed and can be treated effectively.

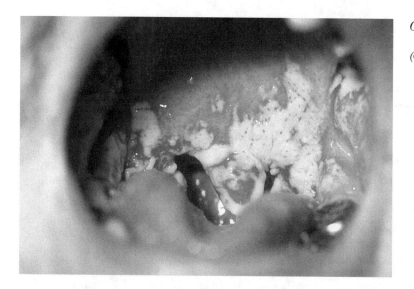

Oral thrush.

(Courtesy CDC)

- **Systemic infections**   These occur when fungi get into the bloodstream and generally cause more serious diseases. Systemic fungal infections may be caused either by an opportunistic organism that attacks a person with a weakened immune system, or by an invasive organism that is common in a specific geographic area, such as cocci and histoplasma. Unlike superficial infections, systemic fungal infections can be life-threatening.

- **Opportunistic infections**   As previously noted, the fungi attack people with weakened immune systems. These can be either systemic or superficial infections.

Several kinds of fungal infections are described in the following sections.

Potent Fact

The fungi that cause infections in people are divided into three groups: yeasts, moulds, and dermatophytes. Only about 180 of the 250,000 known fungal species are able to cause disease. Most of them are moulds, but there are also many disease-causing yeasts. Some fungi are highly pathogenic and can establish a systemic infection in exposed individuals. Others only cause disease when the immune system is weak.

Candida Infection

Candida is a yeast that is common in people—hundreds of thousands of them live peacefully in our bodies. They are found on the mucosal surfaces of the mouth, the gut, and the female reproductive system. Most Candida infections occur on one of those three surfaces. These infections can usually be treated with antifungal drugs, but drug resistance is becoming more and more common, making these infections tougher to cure. Candida can also cause more serious systemic infections.

These infections are opportunistic and occur when the environment in the body becomes favorable for the organism to grow and spread. This can happen, for example, when someone is being treated with antibiotics. In this case, fungi, which compete for nutrients with bacteria within the body, proliferate once the bacteria are eliminated.

> **Potent Fact**
>
> Candida vaginitis, commonly called "yeast infection," is a problem that 75 percent of women will experience in their lifetime. Nearly half of all college-age women will have had at least one episode of a yeast infection.

Yeasts in the body like Candida are usually contained through competition with other microorganisms and the action of host defense systems, such as an intact skin, salivary secretions, and antibody and cell-mediated immunity. Yet, suppression of these systems by antibiotic therapy and or by disease related immune modulation gives this organism sufficient competitive advantage so that it can dominate. This leads primarily to superficial disease but may also result in invasive fungal disease.

Fungal Meningitis

Fungal meningitis, an infection that causes inflammation of the membranes covering the brain and spinal cord, is one of the most common life-threatening opportunistic infections of HIV patients. It is caused by *Cryptococcus*, a fungus that has been found throughout the world. Infections in people are acquired by inhalation of small fungal cells spread on air currents. The organisms survive and spread only in people with weakened immune systems.

The symptoms of fungal meningitis differ from one patient to another, but can include headaches, drowsiness, and confusion. In order to differentiate fungal meningitis from bacterial and viral meningitis, blood and spinal fluid must be tested. The diagnosis is fast if fungal cells are seen in cerebrospinal fluid examined under a microscope. There is also a rapid blood test for blood and spinal fluid that gives positive results in over 90 percent of cases.

> **Antigen Alert**
>
> Beware of white statues. Pigeon droppings are a major source of *Cryptococcus* because the organism grows very well in the nitrogen rich excrement.

The frequency of fungal meningitis has been decreasing in the United States and Europe since the mid-1990s due to the development of more effective antiretroviral therapy for AIDS patients and preventive treatment regimens. Successful treatment of invasive disease involves various antifungal drugs, most given intravenously, but treatment failures are common due to drug resistance. In non-AIDS patients, drug therapy lasts 6 to 10 weeks. Most AIDS patients relapse within six months after treatment ends, so they are usually given lifelong maintenance treatment with an oral antifungal drug.

Aspergillis

Aspergillis is a mould that spreads through the air and can cause serious pulmonary and bloodstream infections in people with weakened immune systems. It is common in cancer patients, bone marrow transplant patients, and people with HIV. Early diagnosis is critical for successful treatment, although mortality is generally quite high. *Aspergillis* accounts for 30 percent of fungal infections among cancer patients and 10 to 25 percent in leukemia patients.

The disease is tough to diagnose because it often occurs in people whose immune systems are deficient, and those people don't show the usual signs and symptoms that would lead the doctor to look for a fungal infection. Also, more invasive diagnostic techniques, such as biopsies, that would be used in healthy people are more dangerous to someone with an immune system problem.

Potent Fact

Aspergillis moulds often live in air conditioning systems in hospitals. Hospitals need to have procedures in place to reduce the risk of spreading these moulds. *Aspergillis* can act as a powerful allergen, often resulting in *Aspergillis* asthma.

The best way to deal with *Aspergillis* infection is to prevent it, but that's tough because the spores are found everywhere—in the soil, the air, and the water! There are spores in your house—in the basement, bedding, air conditioning, potted plants, wicker, and dust.

PCP: Pneumocystis Pneumonia

PCP was one of the first unusual infections observed early in the AIDS epidemic and suggested that patients had weakened immune systems. Approximately 60 percent of people are infected with pneumocystic fungi by the time they are four years old. After a mild infection in childhood, the organism remains latent until weakened immunity triggers a reactivation. The organism lives in soil and water.

PCP has a 40-day incubation period in AIDS patients, and the disease may be accompanied by weight loss, malaise, diarrhea, dry cough, shortness of breath, and low-grade fever. In other types of patients, the onset is subtle, with an average incubation period of 60 days and resulting in a dry cough that can progress to more severe respiratory distress. Besides AIDS patients, people at highest risk for the disease include infants with severe malnutrition, children with primary immune system deficiencies, patients with cancer, and transplant patients.

Patients at highest risk for PCP also include infants with severe malnutrition, children with immune system deficiencies, and recipients of organ or bone marrow transplants. PCP is treatable, and patients at high risk for it can take preventive medicine. Early detection and treatment are the key to beating the illness.

Fungal Infections of the Skin

Fungi that commonly cause skin diseases are called dermatophytes. "Dermatophytes" doesn't refer to a particular group of fungi, but rather to the fact that they attack the dermis, or skin. Fungal infections of the skin can be treated with topical creams as well as prescription drugs.

Athlete's Foot

The best-known fungal skin infection is athlete's foot. It infects approximately 10 percent of the United States population. It is most common among adolescents and adults; however, it may affect people of any age.

Athlete's foot can grow on the feet in different forms, including the following:

◆ **Interdigital:** Infection occurs between the toes, with scaling, fissuring, or softened skin.

◆ **Moccasin:** The fungi grows as a thick scaling over the entire sole of the foot (like a moccasin) and causes discomfort.

◆ **Vesicular:** The fungi appear as small, itchy blisters near the instep.

◆ **Ulcerative:** The infection involves peeling, oozing discharge, and a strong odor that usually starts as red, itchy swelling between the toes.

A good way to combat athlete's foot is to keep feet clean and dry. Topical powders or creams may also help to control infection. Unfortunately, athlete's foot is tough to eliminate and often comes back.

Scalp Itch

Scalp itch is a fungal infection of the scalp and hair. It usually occurs in young children, but may appear in all age groups. It is contagious and may be spread from child to child in a school or day care setting.

An antifungal drug called riseofulvin cures scalp itch in one to three months.

Nail Fungus

Nail fungus is most common in adolescents and adults, especially among people who have frequent manicures. These infections can manifest themselves in a variety of patterns. Sometimes a portion of the nail becomes thick and brittle. Other times, the fungi attack the cuticle and the growth spreads out from there. This cuticle-based infection is common in AIDS patients.

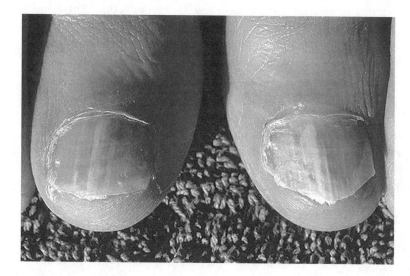

A fungal infection of the toenail.

(Courtesy CDC/Dr. Edwin P. Ewing, Jr.)

Fighting Fungus

As noted earlier in this chapter, the strength or weakness of the body's immune system is one of the most important factors in whether a fungus will cause disease and how severe that disease will be. A person with a healthy immune system is much less likely to get a systemic fungal infection than someone whose immune system is compromised due to disease or ongoing medical treatment.

The primary barriers we have against fungal infections are the same as those for other infections: intact skin, naturally occurring chemicals produced by our body, competition with normal bacteria that live in and on us, and the turnover rate of our skin cells (see Chapter 4 for more on how the body naturally fights disease). Our body even produces its own antifungal substance, which is secreted by our mucous membranes.

Antigen Alert

Women who take systemic antibiotics for treating bacterial infections are more likely to experience opportunistic yeast infections.

In most cases, fungi only cause disease if people have an underlying immune system problem. Most fungi that do cause disease live with us, like Candida, or around us, like *Aspergillis* and *Cryptococcus*. Minimizing exposure to fungi is important for hospitalized patients, but it is difficult to avoid these ubiquitous organisms.

First You Gotta Find It

It's not always easy for doctors to determine if a patient has a fungal infection. Physicians can grow cultures from a patient's skin scrapings or mucosal discharge from the throat or nose. X-rays and CT-scans can be helpful, and doctors sometimes identify fungi by examining tissue samples under a microscope.

Getting Rid of Fungi

Once a person has been diagnosed with a fungal infection, it can usually be treated with an antifungal drug or cream. Some are even available over the counter, such as treatments for vaginal yeast infections, but they should be used carefully and in consultation with a doctor to ensure they are the right treatment.

Antifungal drugs work by inhibiting the growth of a particular chemical in the cell membrane of fungi. Because this chemical is different from chemicals in the human cell wall, the antifungal drugs attack only the invading fungi, leaving the body's own cells intact. Other drugs used against fungi inhibit DNA and RNA replication, preventing the cells from multiplying.

> **CAUTION**
>
> **Antigen Alert** _____
>
> Numerous topical antifungal drugs are available to treat vaginal yeast infections, and the availability of over-the-counter products for self-treatment has made the process easier. However, there is some concern that frequent use of these products (more than twice a year) may contribute to the emergence of drug-resistant Candida that can jeopardize a woman's health. Preliminary studies show only a low level of resistance, but the jury is still out and caution is recommended when using these products. It's still best to see your doctor to get a proper diagnosis and appropriate treatment.

The most widely-used antifungal drug, amphotericin B, is considered the gold standard because it is able to kill many different types of fungi. In the United States, this drug is only available intravenously.

Another class of drugs, called azoles, prevent new cell growth, but don't kill fungal cells that are already there. Most of the azoles can be used topically and are common ingredients in Mycelex and Monistat, used to treat yeast infections.

The Least You Need to Know

◆ Only a few of the thousands of species of fungi are harmful to humans.

◆ Fungi often live peacefully in the body or the environment, only causing infection if the immune system is weakened.

◆ Superficial fungal infections only affect the skin and mucous membranes and usually cause only mild infections.

◆ Systemic fungal infections can invade the body, including the organs, and can cause severe illness or death.

13

Watery Deaths: Cholera and Dysentery

In This Chapter

- ◆ Major water-borne diseases
- ◆ Why water-borne disease primarily affects impoverished areas
- ◆ Causes, symptoms, diagnosis, and treatment
- ◆ Preventing the spread of cholera and dysentery

Water-borne diseases are caused by water that has been contaminated by human or animal wastes, and include diseases such as cholera, typhoid, shigella, polio, meningitis, and hepatitis A and E.

Humans can act as hosts to the bacterial, viral, or protozoal organisms that cause these diseases. In many countries where sewage treatment is inadequate, human wastes are disposed of in open latrines, ditches, and canals or are spread on cropland, resulting in extensive diarrheal disease. It is estimated that 4 billion cases of diarrheal disease occur every year, causing 3 million to 4 million deaths, mostly among children. Worldwide, the lack of sanitary waste disposal and of clean water for drinking, cooking, and washing is to blame for over 12 million deaths a year.

Cholera: Scourge of the Poor

Cholera is not a problem in developed countries, but it is a major public health problem for developing countries, where outbreaks occur seasonally and are associated with poverty and poor sanitation. The disease causes profuse watery diarrhea that leads to rapid dehydration and—if not treated—death. Serious disasters, such as hurricanes, typhoons, or earthquakes, cause a disruption in water systems resulting in the mixing of drinking and waste waters, which increase the risk of contracting cholera among area residents.

Drinking water can be contaminated during natural disasters such as floods, earthquakes, and hurricanes.

(Courtesy CDC)

Compared to diseases like smallpox and tuberculosis, which have been around for thousands of years, cholera is a relatively new disease. Its origins can be traced to India in 1826, but by 1830, 40,000 people a year were dying from cholera. In 1831, nearly 200,000 Russians died. The same year, cholera spread to Poland, Hungary, and Germany, killing hundreds of thousands. As it spread throughout Europe, the death toll rose dramatically. In 1848, Russia alone suffered 3 million deaths! Large outbreaks continued for many decades, killing hundreds of thousands of Europeans each year.

Cholera spread throughout the world in seven large pandemics. The seventh pandemic began in 1961 in Indonesia and subsequently affected approximately 100 more countries. In some areas, more than 20 percent of the people got sick and often half of them died because medical treatment was not available. The pandemic reached Africa in 1970 and moved rapidly throughout the region. By the end of 1971, 25 African countries were reporting cholera outbreaks. Between 3,000 and 43,000 cholera cases were reported in Africa every year until 1990. The following year, a large epidemic affected 14 countries and resulted in more than 100,000 cases and 10,000 deaths.

> **Infectious Knowledge**
>
> His obsession was the dangerous lack of sanitation in the city. He appealed to the highest authorities to fill in the Spanish sewers that were immense breeding grounds for rats, and to build in their place a closed sewage system whose contents would not empty into the cove at the market, as has always been the case, but into some distant drainage area instead. The well-equipped colonial houses had latrines with septic tanks, but two thirds of the population lived in shanties at the edge of the swamps and relieved themselves in the open air. The excrement dried in the sun, turned to dust, and was inhaled by everyone along with the joys of Christmas in the cool, gentle breezes of December ... He was aware of the mortal threat of the drinking water.
>
> —Gabriel Garcia Marquez, *Love in the Time of Cholera*

Cholera's Causes

Cholera is caused by a bacterium that has two specific toxins, or poisons. One causes profuse diarrhea, and the other assists the bacteria in living and multiplying in the intestines.

Most people infected develop mild cases or do not get sick at all, although the bacteria may live in their gut for 7 to 14 days. The vast majority of people who get cholera are cured when rapid treatment is available. However, without treatment, death rates are between 50 and 70 percent.

> **CAUTION**
>
> **Antigen Alert** _____
>
> The disease-causing genes in the bacterium that causes cholera are on one particular chromosome, or portion of the bacteria's DNA. Sometimes these genes are transferred from the bacteria to other organisms. When this happens, the organism that receives the piece of DNA from the cholera-causing bacteria goes on to cause disease.
>
> A variety of bacteria exchange genes this way. It can happen through direct physical contact between organisms, uptake of free DNA, or when certain viruses, called bacteriophages, carry genetic material from one organism to another.
>
> In addition to making some bacteria virulent, genetic transfer can also help spread drug resistance genes from one organism to another.

Don't Drink the Water

Cholera is a diarrheal disease caused by the bacterium *Vibrio cholerae*. It is almost always transmitted by water or food that has been contaminated by human waste. Surface water and water from shallow wells are common sources of infection, where bathing or washing cooking utensils in contaminated water can also transmit cholera. Foods such as cooked grains that are allowed to sit at or above room temperature for several hours can also be an excellent breeding environment. Raw or undercooked shellfish and raw fruits and vegetables can also transmit the bacterium.

People with infections but no symptoms can carry the disease from place to place causing it to spread, although person-to-person contact is not a significant mode of transmission.

Corralling Cholera

Positive diagnosis of cholera requires identification of the bacteria from stool specimens. Fortunately, cholera can be successfully treated by rapid oral or intravenous fluid and electrolyte replacement. Antibiotic therapy can shorten the duration of the disease but does not affect the severity of an attack.

Areas without a safe water supply and good sanitation are at most risk for epidemic cholera. The key to preventing its spread is limiting the growth and survival of the organism that causes it. Outbreaks can be minimized by educating the public about food and water safety, the importance of hand-washing, and the need to use toilets. When an epidemic of cholera occurs, however, total cases and deaths can be reduced by early detection and rapid initiation of treatment and control methods. Cholera epidemics are unpredictable and may recur in either the rainy or dry season. Fortunately, immunity gained from having an infection before protects a person against cholera reinfection.

> **Potent Fact**
>
> The cholera bacterium grows poorly in an acidic environment, so people taking antacids or other products to reduce stomach acid are more susceptible.

Protecting Against Cholera

The only cholera vaccine available now gives partial protection for only three to six months—it does not provide long-term protection like other vaccines. For that reason, it is not recommended to prevent or control cholera outbreaks because it may give a false sense of security to those vaccinated and to public health authorities, who may then forgo implementing more effective measures.

There are two oral vaccines with few side effects that are given outside the United States. These vaccines provide 60 to 100 percent protection against major outbreak strains for at least six months.

Epidemic Dysentery

Dysentery is an inflammation of the intestine characterized by the frequent passage of feces with blood and mucus. Like cholera, dysentery is spread by fecal contamination of food and water, usually in impoverished areas with poor sanitation. Epidemics are common in these areas. A four-year epidemic in Central America, starting in 1968, resulted in more than 500,000 cases and more than 20,000 deaths. Since 1991, dysentery epidemics have occurred in eight countries in southern Africa (Angola, Burundi, Malawi, Mozambique, Rwanda, Tanzania, Zaire, and Zambia).

Epidemic dysentery is a major problem among refugee populations, where overcrowding and poor sanitation facilitate transmission. Epidemics are characterized by severe disease, high death rates, person-to-person spread, and multiple antibiotic resistance. Worldwide, approximately 140 million people develop dysentery each year, and about 600,000 die. Most of these deaths occur in developing countries among children under age five. In the United States, only about 25,000 to 30,000 cases occur each year.

Antigen Alert

In sub-Saharan Africa, diarrheal diseases are a leading cause of death in children under age five. It is estimated that each child has five episodes of diarrhea per year and that 800,000 of those children will die from diarrhea and associated dehydration.

Dysentery's Roots

Dysentery is most commonly caused by one of two different organisms: One is a bacterium called *Shigella;* the other is caused by an amoeba. *Shigella* is the most important cause of bloody diarrhea because it destroys cells that line the large intestine, which leads to mucosal ulcers in the intestine. The mucosal ulcers cause the bloody diarrhea. Ingesting as few as 10 to 100 bacteria, which can be contained in a tiny amount of infected food or water, can cause disease.

Amoebic dysentery is prevalent in regions where human excrement is used as fertilizer. The amoebas that cause dysentery can form cysts, which are like bacterial spores that can become inactive and highly resistant to environmental conditions. In other words, they can live a long time outside the body and then reactivate and cause disease when conditions become favorable.

Potent Fact

The two primary causes of dysentery are the *Shigella* bacterium and an amoeba. *Shigella* can cause severe disease and epidemics, although it responds well to treatment. Dysentery caused by the amoeba is milder than its bacterial cousin, although it is quite difficult to treat and cure and often becomes chronic.

Cysts and live amoebas are excreted in the feces of an infected person, but only the cysts can survive outside the body. The amoebic infection is milder in comparison with bacterial dysentery. Despite this, amoebic dysentery is more difficult to treat and cure; bacterial dysentery responds better and more quickly to treatment.

Both types of dysentery infect people of diverse age, sex, and ethnic backgrounds, although children are more susceptible.

Symptoms

Patients with bacterial dysentery often have fever, abdominal cramps, rectal pain, and bloody stools. Occasionally, large portions of the intestinal membrane pass with particularly foul-smelling stool containing yellowish white mucus and/or blood. In nearly half the cases, *Shigella* does not cause bloody diarrhea.

Disease Diction

The amoebic form of dysentery lives outside the body by forming **cysts,** which are similar to bacterial spores. Cysts have a tough outer wall that prevents different environmental conditions from killing the amoebas. Similar to a bear hibernating, the cysts are dormant until conditions are better—they invade a body that has the right temperature and nutrients—then they wake up and cause disease.

When amoeba cysts are ingested with contaminated food or water, they germinate and develop into live amoebas in the intestine. The disease remains mild if the amoeba stay confined within the intestines. Like bacterial dysentery, invasion of the intestinal wall leads to fever, abdominal and rectal pain, and bloody diarrhea. Amoebic dysentery may occur in a chronic form when the amoebas invade blood vessels of the intestine and are carried to other parts of the body, causing amoebic abscesses of the liver and brain. About 40 percent of all untreated cases eventually cause nonintestinal infections, such as amoebic hepatitis.

Diagnosing Dysentery

Dysentery is diagnosed from rectal swabs that show evidence of dysentery-causing *Shigella* bacteria or amoeba.

Disposing of Dysentery

Bacterial dysentery often subsides by itself, although treatment using antibiotics is recommended to prevent recurrence. Having drug-susceptibility tests performed before beginning treatment is important to determine which antibiotics will work best, because many organisms have become drug resistant. *Shigella* first began to acquire resistance in the 1940s and has become resistant to several classes of drugs since then.

Treating the dehydration that accompanies dysentery is also important. These symptoms should be treated with oral rehydration salts or, if severe, with intravenous fluids.

A combination of drugs is used to treat amoebic dysentery: an amoebicide to eradicate the organism from the intestinal tract, and an antibiotic to eliminate potential secondary bacterial infections.

Eradicating the Epidemic

Early detection and notification of epidemic dysentery, especially among adults, allows for speedy reaction to help fight the disease's spread. Hand-washing with soap and water can reduce secondary transmission of *Shigella* infections among household members. And among larger groups, such as within refugee camps, the most effective strategies to control transmission of epidemic *Shigella* are to ...

- ◆ Distribute soap.
- ◆ Provide clean water.
- ◆ Promote hand-washing before eating or preparing food and after defecation.
- ◆ Install and maintain proper sewage systems or treatment facilities.

Developing a Vaccine

There are no vaccines for dysentery, although there is a strong need, particularly because drug resistance often limits treatment options. There are currently several potential vaccines in the evaluation stages.

Contaminated water causes millions and millions of cases of disease every year. We have discussed cholera and dysentery in this chapter, but there are other diseases, too. Improved sanitation is key to controlling these diseases, but until conditions improve, it is important for victims to receive proper treatment and to be sure to prevent the severe dehydration that often occurs with diarrheal disease.

The Least You Need to Know

- ◆ Cholera and dysentery are major water-borne diseases that are common in less developed parts of the world.
- ◆ The bacterial form of dysentery is more likely to cause epidemics and severe disease.
- ◆ Dehydration is a major symptom of cholera and dysentery that may need special treatment.
- ◆ Improved sanitary conditions can help to control the spread of both diseases.

Mosquito and Tick-Borne Diseases

In This Chapter

- Diseases caused by mosquitoes and ticks
- Ways to protect against bug-borne diseases
- How likely it is you'll be infected

As you already learned in Chapter 9, mosquitoes serve as vectors, or carriers, of the malaria parasite. Bloodthirsty mosquitoes and ticks also carry disease-causing bacteria and viruses, which can be passed on to people when they are bitten by the bugs. And unlike malaria, which isn't endemic to the United States, these other mosquito- and tick-borne diseases are sometimes found in our own backyards.

This chapter covers three mosquito-borne diseases and two tick-borne diseases. The first two, yellow fever and dengue, are also called hemorrhagic fevers, because one of their symptoms is small hemorrhages under the skin. Diseases like Lyme are spread by ticks, while West Nile is spread by mosquitoes. West Nile doesn't cause many infections but it has caused a great deal of panic due to heavy media attention in the United States.

Yellow Fever

Yellow fever is caused by a virus that is spread by several species of Aedes and Haemogus mosquitoes. Infection causes disease, with symptoms ranging from a mild flu to severe illness, including high fever, severe headache, muscle pain, jaundice, and vomiting of a liquefied black putrid matter. It is the pronounced yellow skin and eye color of severe jaundice that gives rise to the name yellow fever. Historically, epidemics occur when the disease breaks out in previously unexposed populations.

Yellow fever and dengue are caused by the same types of mosquitoes.

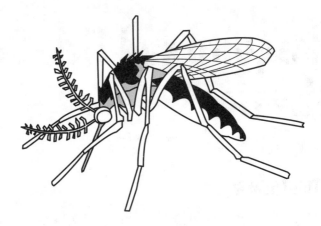

The natural sources of yellow fever are monkeys and mosquitoes, which inhabit the jungles of South America and Africa. Previously unexposed Europeans who traveled to the Caribbean, Central America, and North and South America were common victims for several hundred years. In 1741, 20,000 British soldiers died of the disease, then called "Black Vomit," during an expedition to capture Peru and Mexico.

Disease Diction

Yellow fever got its name because of the severe jaundice that is a symptom of the disease—it causes pronounced yellow skin and eye color. At one time it was called "Black Vomit" because of the characteristic black putrid vomit it produces.

Yellow fever was brought to the Americas on slave ships in the 1500s. However, it was United States coastal towns that were particularly vulnerable to the disease in the seventeenth and eighteenth centuries. In 1793, the city of Philadelphia lost nearly 10 percent of its population to an epidemic of yellow fever. Eighty-five years later, in 1878, 20,000 people in the United States died from the disease. The last major United States epidemic occurred in New Orleans in 1905, and the last urban case of the disease occurred in 1942. However, in an all-too-familiar scenario, two cases were reported in 1996 and 1999 from returning foreign travelers.

There are approximately 200,000 cases of yellow fever, resulting in 30,000 deaths, each year in tropical endemic areas of Africa and the Americas. Thirty-three countries, with a combined population of 508 million, are at risk in Africa. In the Americas, yellow fever is endemic in several Caribbean islands and in nine South American countries, including Bolivia, Brazil, Colombia, Ecuador, and Peru, where sporadic infections occur in forestry and agricultural workers. In Africa, the virus is transmitted in the damp savanna zones of West Africa during the rainy season, and infections occur principally among children. Periodically, large outbreaks occur in cities and villages, resulting in many thousands of cases.

Despite an effective vaccine, the number of yellow fever cases continues to grow worldwide. This is largely due to changes in the world's ecology from deforestation and urbanization, which have increased the mosquito/virus contact.

> ### Infectious Knowledge
>
> Everybody who can, is fleeing from the city, and the panic of the country people is likely to add famine to the disease.
>
> —Thomas Jefferson, commenting on the panic caused by the 1793 yellow fever outbreak in Philadelphia

> ### Infectious Knowledge
>
> Philadelphia was the most prominent American city in 1793, serving as the seat of the federal government. In August of that year, a deadly epidemic emerged. Numerous individuals shared common symptoms: severe fever, nausea, skin eruptions, black vomit, profound lethargy, rapid but weak pulse, incontinence, and a distinct yellow skin and eye color. Yellow fever was spreading through Philadelphia.
>
> Terror gripped the city as nearly 600 people died of the disease in the first four weeks. Half the population fled in panic. Most business and commerce ceased. All federal, state, and municipal government was suspended, and President George Washington abruptly left the city. By the time the epidemic waned some three months later, 5,000 people, or nearly 10 percent of the population, had succumbed to yellow fever.

Yellow Fever's Virus

Yellow fever is caused by a virus, which infects monkeys as well as humans. The mosquito that carries the virus is most active in the early morning and late afternoon. It breeds in fresh water in both natural and artificial containers in and around human dwellings, including small ponds, gutters, old tires, flowerpots, and water storage containers.

Putting the "Yellow" in Yellow Fever

Once someone is infected, the virus remains silent in the body during an incubation period of three to six days. There are then two disease phases. An "acute" phase is characterized by fever, chills, muscle pain, backache, headache, loss of appetite, and nausea. After three to four days, most patients improve and their symptoms disappear.

However, some 15 percent of infected patients enter a second and more deadly "toxic" phase. These patients experience a reappearance of a fever, pronounced jaundice occurs resulting in distinct yellowing of the skin and eyes, abdominal pain, and vomiting. Bleeding may occur from the mouth, nose, and eyes. Bleeding within the stomach produces the so-called "black vomit." As kidney function deteriorates, patients often die within 10 to 14 days in the "toxic phase."

Recognizing Yellow Fever

Yellow fever is difficult to recognize and in its early stages can be mistaken for malaria, typhoid, rickettsial diseases, other hemorrhagic viral fevers, or viral hepatitis. Specific diagnosis depends on isolation of the virus from blood and/or finding viral antigen in a sample. Liver biopsy, or surgically removing a portion of the liver for testing, has been used to confirm the diagnosis, but it is not recommended given the hemorrhagic properties of the disease.

How You Catch It

Humans and monkeys are the primary animals infected by the virus. The virus spreads from one animal to another by the mosquito. Infected mosquito eggs can lie dormant through dry conditions and hatch when the rainy season begins. The transmission cycles for yellow fever include "jungle," "intermediate," and "urban." "Jungle" yellow fever occurs in tropical rainforests where infected monkeys pass the virus on to wild mosquitoes that feed on them. The mosquitoes in turn bite humans entering the forest, resulting in sporadic cases of yellow fever. "Intermediate" yellow fever causes small epidemics in humid or semi-humid savannahs of Africa. "Urban" yellow fever causes epidemics in cities and villages with large numbers of cases.

Breaking the Fever

There is no specific treatment for yellow fever. Typically, patients are treated for dehydration and fever with rehydration salts. Intensive care is needed to improve the outcome for seriously ill patients.

Keeping the Mosquitoes at Bay

Mosquito-control measures to reduce breeding sites are important to prevent virus transmission. Similarly, general precautions to avoid mosquito bites are effective at preventing bites and infection. These methods include the use of insect repellent, protective clothing, and mosquito netting. In many endemic areas, spraying is impractical and vaccine programs are the best way to limit disease outbreaks. The World Health Organization (WHO) strongly recommends routine childhood vaccination to prevent an epidemic in endemic countries.

Tried and True Protection: Vaccination

Vaccination is the single most effective way to prevent yellow fever. The vaccine has been used since the 1930s and is safe and highly effective in preventing yellow fever in adults and children over nine months of age. A single dose of vaccine confers immunity within one week and lasts for 10 years or more. Minor symptoms like mild headache and muscle pain occur within 10 days of vaccination in less than 5 percent of vaccinated individuals.

More serious side effects are extremely rare but medical help should be sought in the event of seizures, difficulty breathing or swallowing, fast heartbeat, feeling of burning, tingling of skin, severe headache, skin rash or itching, sneezing, stiff neck, throbbing in the ears, unusual tiredness or weakness, and/or vomiting. The vaccine should not be given to people with weakened immune systems, pregnant women, or people who are allergic to eggs.

Potent Fact

Vaccination is highly recommended for people traveling to high-risk areas. A vaccination certificate is required for entry to many countries in Asia, Africa, and South America.

Dengue

Dengue hemorrhagic fever (DHF) and dengue shock syndrome (DSS) occur in more than 100 countries and territories. The disease is endemic in Africa, the Americas, Eastern Mediterranean, and most prominent in South East Asia and the Western Pacific. In the last 25 years, the number of countries with DHF has increased more than fourfold. In 1998, 1.2 million cases of dengue and DHF were reported to WHO, accounting for 15,000 deaths.

The number of reported cases is believed to represent only a very small percentage of the global disease, which by some accounts may top 51 million infections each year and has the potential to impact two fifths of the world's population.

Dengue hemorrhagic fever causes death in more than 20 percent of cases of the disease when medical attention is unavailable. That percentage can be reduced to less than 1 percent, however, with modern medical intervention. The same mosquito that carries the yellow fever virus can carry the dengue virus as well. The global rise in urban populations is bringing large numbers of people into contact with infected mosquitoes.

What Causes Dengue

Dengue is a virus that is closely related to the yellow fever virus. Dengue, however, comes in four "flavors," with the different types causing different forms of disease, ranging from mild to life-threatening. The mosquito that carries it breeds in fresh water in both natural and artificial containers, in and around human dwellings, wherever standing water may accumulate.

Where the mosquitoes that cause dengue can breed.

(Courtesy PAHO)

The variety of breeding places of the Dengue mosquito in your surroundings
Feti Dengue

1. Old tyres
2. Laundry tanks
3. Uncovered tanks
4. Drums/Barrels
5. Discarded buckets and other containers
6. Pet dishes
7. Construction blocks
8. Bottles
9. Discarded tin cans
10. Tree holes & bamboos
11. Bottle pieces on top of walls
12. Old shoes
13. Flower pots & saucers
14. Discarded toys
15. Roof guttering
16. Bromeliad plants
17. Garden containers & tools
18. Brick holes

Without containers there is no mosquito; without mosquitoes there is no Dengue.
Get rid of breeding places in your surroundings.

What Dengue Feels Like

Dengue virus infection may cause a mild illness, dengue fever, or dengue hemorrhagic fever, including dengue shock syndrome. The severity of the disease depends on age, immune status, and the virus strain. Infants and children usually develop mild symptoms with fever similar to other viral infections. Most cases of dengue are benign and end after about seven days.

Dengue fever in older children and adults is characterized by a sudden onset of high fever, severe headache, muscle aches, intense muscle and joint pain (break-bone fever), nausea, weakness, vomiting, and rash. A measles-like rash shows up three to four days after the start of the symptoms and begins on the torso, spreading out to the face, arms, and legs. Fever and rash may subside, and then reoccur. All these symptoms usually persist for about seven days. The liver may also be enlarged in 10 percent of children.

The hemorrhagic form of dengue fever is much more severe and is associated with loss of appetite, vomiting, high fever, headache, and abdominal pain. Bleeding develops in the gums, skin, and intestinal tract, leading to a sudden drop in blood pressure with the onset of shock, circulatory failure, and in some cases, death.

Are You Likely to Get It?

People infected with one type of virus maintain lifelong immunity to that type, but remain susceptible to infection with other strains of dengue. Severe disease is more likely to develop if a person was previously infected with one type, or was born to a dengue-immune mother.

Four different dengue viruses have been implicated in both dengue fever and dengue hemorrhagic fever. Dengue hemorrhagic fever occurs when the patient contracts a different dengue virus after previous infection(s) by another type. Prior exposure to a different dengue virus type actually makes people more susceptible to dengue hemorrhagic fever and dengue shock syndrome.

Worldwide, over 100 million cases of dengue fever occur every year. A small percent develop into dengue hemorrhagic fever. Most cases in the United States are brought in from other countries. It is possible to pass the infection from a traveler in the United States to someone who has not traveled.

Antigen Alert

Risk factors for dengue hemorrhagic fever include having antibodies to dengue virus from prior infection, age less than 12 years old, being female, being Caucasian, and not having good nutrition.

Determining If You Have It

A positive diagnosis can be made with a blood test that looks for an antibody response to infection. Antibodies can be detected within six or seven days of onset of illness.

Waiting Out the Disease

There is no specific treatment for dengue fever. Typically, patients are given fluids and acetaminophen for fever and pain. Hospitalization is required if hemorrhagic fever or shock develops. Careful clinical management by health professionals can save the lives of dengue hemorrhagic fever patients.

The Catch-22 of Immunization

Because protection against only one dengue virus may actually increase the risk of the more serious disease, developing of a vaccine for dengue is very complicated and difficult. Right now, there is no vaccine available.

The only effective method of controlling or preventing dengue is to reduce the population of mosquitoes—vectors—that carry it. Vector control involves using environmental management and pesticides on larval habitats. During outbreaks, emergency control measures may also include widespread spraying of insecticide to kill adult mosquitoes. However, such approaches are short-lived. Regular monitoring and surveillance of the natural mosquito population are important for overall prevention.

West Nile Encephalitis

West Nile encephalitis is another viral infection that is transmitted from mosquitoes to people. The first documented case of the disease in the Western Hemisphere was in the United States in 1999. Although West Nile has caused tremendous worry, it has not caused that many infections. This disease has also been found in Africa, West Africa, Eastern Europe, and the Middle East.

Disease Diction

Encephalitis is swelling of the brain. It can be caused by viruses and bacteria, including those transmitted to people by mosquitoes.

In 1999, there were 62 severe cases of West Nile in the United States. Seven people died. In 2000, there were 21 cases and two deaths in New York City. There are no worldwide statistics available on the incidence of this disease.

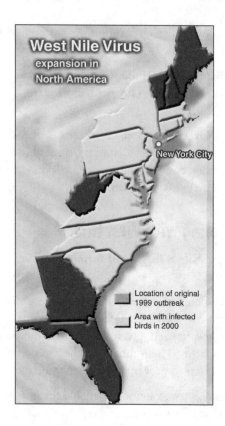

West Nile Virus
expansion in
North America

New York City

Location of original
1999 outbreak

Area with infected
birds in 2000

West Nile virus expansion in North America.

(Courtesy United States Department of Defense)

In the United States, the disease is most likely to strike in the late summer and early fall. In warmer climates, the disease may strike at any time. The highest number of cases of West Nile have been in Florida, where mosquitoes have the opportunity to breed year-round.

From Birds, to Mosquitoes, to the Human Nervous System

West Nile infection occurs when an infected mosquito bites a person. The mosquito gets the virus from feeding on infected birds. The virus lives in the mosquitoes' salivary glands and travels through the human bloodstream to enter the brain and interfere with the central nervous system. Essentially, it causes swelling of the brain tissue.

CAUTION **Antigen Alert**

The appearance of dead birds may be a warning that West Nile is present in an area. Public health officials now test birds for the presence of the virus.

Less than one percent of those infected will develop severe illness. Three to five percent of those with severe illness will die.

Most West Nile infections are mild. Symptoms include fever, headache, and body aches. Sometimes there is a skin rash and swollen glands. The incubation period is three to fifteen days.

Infectious Knowledge

Unlike many infections that are species-specific, West Nile has been seen in birds, horses, cats, bats, chipmunks, skunks, squirrels, and domestic rabbits.

Potent Fact

There is no evidence of the infection being contagious between vertebrate species. In other words, people can't get West Nile from a horse or a cat.

In a severe infection, symptoms are headache, high fever, neck stiffness, stupor, disorientation, coma, tremors, convulsions, muscle weakness, paralysis, and on rare occasions, death.

Dealing with West Nile

West Nile is diagnosed through patient history and blood tests.

Like dengue and yellow fever, there is no specific treatment for West Nile infection. In severe cases, patients are hospitalized, given intravenous fluids, and put on a ventilator to help them breathe. Sometimes antibiotics are given to prevent secondary infection.

There is a vaccination available for horses, but no vaccine yet for people.

Mosquito Management Is Key to Prevention

As with other bug-borne diseases, good mosquito management to track growth in mosquito populations helps to prevent West Nile infections. This also includes using insecticides in high incidence areas. Disease surveillance is also an important component of a good prevention strategy.

Staying indoors at dawn and dust, wearing long-sleeved shirts and long pants, and using bug repellent containing DEET or permethrin can help, too. Bug repellent should be sprayed on clothes and skin.

Army Spc. Clark Dutterer, a preventive medicine specialist at the U.S. Army Center for Health Promotion and Preventive Medicine at Aberdeen Proving Ground, Maryland, sets a light trap to catch adult mosquitoes in the Maryland woods. The mosquitoes are then tested to see if they are carrying certain diseases. The trap was supplied by the Centers for Disease Control and Prevention.

(Photo by Sgt. 1st Class Kathleen T. Rhem, USA. Photo courtesy United States Department of Defense)

Rocky Mountain Spotted Fever

Rocky Mountain spotted fever was first discovered in the Snake River Valley of Idaho in 1896. It was frequently fatal and many people in the region got sick. By 1900, it had been found in Washington, Arizona, and New Mexico.

Today it is found throughout the United States, Canada, Mexico, and South America.

Rocky Mountain spotted fever is caused by a bacterium that is spread to people through tick bites. Symptoms of the disease include fever, headache,

Disease Diction

Rocky Mountain spotted fever was originally called "black measles" because of the appearance of the rash.

and muscle pain, followed by the appearance of a rash. The disease is hard to diagnose in its early stages and can be fatal.

The bacterium that causes Rocky Mountain spotted fever has a complex life cycle that involves ticks and mammals. People are considered accidental hosts and are not involved in the natural transmission cycle of the bacteria.

Rocky Mountain spotted fever remains a serious and potentially life-threatening disease. Three to five percent of the people who get it die from it. However, effective antibiotic treatment has dramatically reduced the impact of this disease.

Infectious Knowledge

Howard T. Rickets was the first to identify the bacteria that causes Rocky Mountain spotted fever. The genus of bacteria, *Rickettsia*, is named after him. Dr. Rickets died of typhus in Mexico in 1910, soon after he finished his work on Rocky Mountain spotted fever. Typhus is caused by a bacteria that is closely related to the one that causes Rocky Mountain spotted fever—it is from the genus *Rickettsia* as well.

Lyme Disease

In 1977, a group of children in and around Lyme, Connecticut, developed similar cases of arthritis. Research showed that this was an infectious disease caused by a bacterium that is transmitted to people via tick bites.

Ticks ingest the bacteria that cause the disease from deer or rodent carriers. Not all species of ticks carry Lyme disease, but any tick bite should be considered suspicious.

Ticks feed by inserting their mouths into skin and slowly sucking blood. It usually takes 36 hours of tick attachment before transmission occurs, so it is very important to check for ticks after being outside, especially in the spring.

Lyme is mostly, but not exclusively, found in the Northeast United States. There are around 16,000 cases of the disease in the United States each year. Lyme is the most common tick-borne infection in the United States and effects people in almost every state.

People who live or work in places near woods or overgrown brush are at risk.

Hiking, camping, fishing, and hunting can also be risk factors for the disease.

Symptoms of Lyme

After a tick bite where transmission occurs, the person often develops a small red bull's-eye-shaped rash. Other symptoms that follow are fever, fatigue, headache, and muscle and joint aches. The incubation time is seven to fourteen days, but symptoms can show up as quickly as three days after transmission or as long as 30 days later. Some people don't really get many symptoms at all and others experience very generalized symptoms.

If the bacteria get into the nervous system, they can cause meningitis and palsy. If they get into the musculoskeletal system, they can cause severe muscle and joint pain. In the rare instances when they get into the heart, they can cause a variety of blockages. Undiagnosed Lyme can cause joint swelling, sleep disturbance, fatigue, and even personality changes.

Reported Lyme disease cases by state (1999).

Licking Lyme

Diagnosis is based primarily on the appearance of symptoms, particularly if the person knows they've been bitten by a tick. Blood tests help to confirm exposure. Because anti-bodies to Lyme stay in the system for months and even years, the blood tests aren't always accurate and false positives can occur.

Lyme disease is difficult to diagnose. Doctors diagnose Lyme disease based on symptoms, history, and blood test results. If there is a tick bite and a rash, that is strong evidence of infection. However, not all victims get a rash, and not everyone remembers being bitten. Blood tests, also known as Lyme titers, cannot diagnose the disease alone, but they are used to confirm a diagnosis.

The most common tests look for antibodies to the bacteria that causes Lyme disease. However, it takes two to six weeks after infection for those antibodies to be made, so a blood test right after a tick bite may show a false negative. Other bacterial infections may also cause a blood test to be positive when the patient doesn't have Lyme.

To confuse the issue even more, antibodies to the bacteria that causes Lyme disease persist in the body for months and even years after an infection—even if the infection was treated successfully—so the presence of antibodies doesn't always make it clear that an active infection is present.

PCR tests to detect the bacteria itself might work, but they haven't yet been standardized for Lyme disease.

Lyme disease is effectively treated with antibiotics when it is caught early. The treatment usually lasts for three to four weeks. Later stage disease may require intravenous antibiotics for four weeks or more, depending on how bad it is.

Prevention Plays a Key Role

Preventive measures can help reduce the risk of exposure. It is important to avoid infested areas, especially in the spring. Ticks thrive in moist, shaded environments where deer and rodent hosts are abundant.

Wearing light-colored clothing makes it easier to find ticks before they bite. Long-sleeved shirts, tucking pants into boots or socks, and wearing boots that come up above the ankles can also help. Bug sprays that contain DEET or permethrin are also effective ways to prevent tick bites.

> **Potent Fact**
>
> What do you do if you find a tick on your skin? According to the Centers for Disease Control and Prevention (CDC), check yourself daily for ticks. If you find one, use a fine-tipped tweezer to pull it out. Grasp the tick firmly as close to the skin as possible and pull the body away from the skin with a steady motion. Don't use petroleum jelly, matches, or nail polish remover. Don't worry if the mouth parts stay embedded—the bacteria live in the tick's gut.
>
> After removing the tick, cleanse the area with antiseptic. If a rash appears, call your doctor, who may have you take antibiotics as a preventive measure.

A New Vaccine

A vaccine for Lyme disease, called Lymerix, received FDA approval and it was used in the United States for a few years. However, as of February, 2002, the vaccine was taken off the market in the United States. The manufacturer says that demand was very low, but there has also been some controversy about whether the side effects, like arthritis and muscle pain, were too similar to the symptoms of the disease. In other words, some who got the vaccine thought it gave them the disease. Although there was low demand for the vaccine, the number of Lyme cases has not dropped, indicating that there is still a need for a safe and effective Lyme vaccine.

> **Infectious Knowledge**
>
> Lyme disease is one of the infectious diseases that can affect people and animals. Dogs and horses are prone to Lyme.

The Least You Need to Know

◆ Mosquitoes and ticks act as vectors to transmit the diseases from other mammals to people.

◆ Many tick- and mosquito-borne diseases cannot be easily treated and cured. Bug management is the best way to prevent them.

◆ West Nile encephalitis is a devastating disease, but it is quite rare in the Western Hemisphere.

◆ There are vaccines for some of these diseases, but they do not provide full protection.

Chapter 15

Preying on the Young: Childhood Diseases

In This Chapter

- ◆ Why some diseases attack children
- ◆ How vaccinations have saved thousands of young lives
- ◆ The role children play in spreading the common cold
- ◆ The importance of vaccinations today

Diseases can be devastating for anyone, but it seems particularly unfair when they attack children. Unfortunately, many diseases seem to take a special interest in the young, infecting them more frequently and vigorously than they do adults.

In this chapter, you will learn about several common childhood diseases, and why some diseases seem to prey on the young. It's not all gloom and doom, though, for researchers have made great strides in controlling many childhood diseases. Vaccines, in particular, have saved thousands of young lives, and with proper use they will continue to do so.

Why the Young?

Children are more susceptible to diseases for a number of reasons. The major reason for children's increased susceptibility is that they have had limited exposure to diseases and therefore haven't yet built the immunologic defenses required to fend off certain diseases. The environment plays an important role as well. Children in day care centers and in school pass infections around and then take them home and pass them to siblings and parents. This is a cycle that is difficult to break. Children also don't always practice good hygiene and that makes them both susceptible to as well as good transmitters of disease.

The Polio Panic

On a warm August evening in 1921, after spending a relaxing vacation in Campobello Island, New Brunswick, 39-year-old Franklin Delano Roosevelt felt a mild flu coming on. By the morning, FDR had a fever of 102 degrees and had lost strength in his aching legs. As night approached, the pain spread to his neck and back, and soon he couldn't move his legs at all.

The future president of the United States didn't have a case of the flu; he had contracted poliomyelitis, more commonly known as polio, a crippling viral disease that would leave him paralyzed from the waist down for the rest of his life.

Polio has caused sporadic paralysis and death for much of human history. In the twentieth century, epidemics of polio occurred regularly in cities during the summer, leaving large numbers of healthy children and adults crippled or dead due to paralysis of their breathing muscles. The image of polio-stricken individuals confined to an "iron lung," a large metal cylinder that operated like a pair of bellows to allow them to breath, was a frightening reminder of the disease's devastation.

> **Infectious Knowledge**
>
> In the fall of 1952, as the New York Yankees completed yet another World Series victory over the Brooklyn Dodgers, one of the worst polio epidemics in American history was concluding in which 57,628 cases resulted in scores of paralysis and 3,300 deaths.

Generally known as "infantile paralysis," polio infected 27,363 persons in 1916, with more than 7,000 deaths. Each year, the "summer plague" left its mark on a scared public only to vanish with the first frost.

Soon after the introduction of effective vaccines in the late 1950s and early 1960s, polio was brought under control and was largely eliminated as a public health problem in industrialized countries. Today, the disease has been eradicated from large parts of the world. Outbreaks have occurred in Eastern Europe, but the only main remaining major reservoirs of virus transmission are in South Asia and sub-Saharan Africa.

Picking Polio Out from the Crowd

Polio is caused by a virus that enters the body through the mouth and multiplies inside the throat and intestines. The incubation period is usually 6 to 20 days, although it may range from 3 to 35 days. The initial symptoms include fever, fatigue, headaches, diarrhea, vomiting, constipation, stiffness in the neck, and pain in the limbs. The virus persists in the throat for about a week, and it may be excreted—still infectious—in the feces for many weeks.

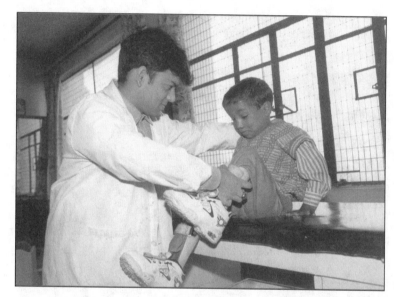

Polio victim in India.

(© Marcel Crozet/WHO)

In 95 percent of people, the infection causes no symptoms. However, in some cases, the virus enters the central nervous system. As it multiplies, the virus destroys nerve cells that activate muscles. The muscles of the legs are affected more often than the arm muscles. In the most severe cases, poliovirus attacks the motor neurons of the brain stem, reducing breathing capacity and causing difficulty in swallowing and speaking. Without life support, death ensues.

Children Are Targets

Polio can strike at any age, but affects mainly children under three. Patients are most infectious from 7 to 10 days before and after the onset of symptoms, but remain contagious as long as the virus is present in the throat and feces.

Unfortunately, there are no effective drugs to treat polio. Typically, the disease is treated by applying moist heat, coupled with physical therapy to stimulate the muscles.

A Most Welcome Vaccine

In the late 1950s, Albert Sabin and Jonas Salk developed effective vaccines against the virus. The Salk vaccine is an injection of chemically killed virus, which primes the immune system to recognize the virus and eliminate it. This vaccine confers lasting, but not always lifelong, immunity. The Sabin vaccine is given orally and contains attenuated, live viruses. The attenuated, or weakened, virus is not strong enough to cause paralysis. It multiplies in the intestinal tract and induces immunity.

Widespread immunization campaigns rapidly brought polio under control in developed countries. A global campaign to eradicate polio, rivaling smallpox, has a real chance for success.

Child being immunized in New Delhi railway station.

(© Marcel Crozet/WHO)

Infectious Knowledge

In 1985, Rotary International launched a program called PolioPlus, with the goal of eliminating polio in the world by 2005. This 20-year commitment began with providing vaccines and has grown to assisting in the field, training laboratory personnel, and working with governments around the world. Rotary has partnered with the World Health Organization in this ambitious endeavor.

They have contributed more than $353 million to date and helped to protect two billion children. By 2005, they will have contributed $500 million.

A Pox on the Young: Chickenpox

More than 95 percent of Americans get chickenpox by the time they reach adulthood. The disease, which is caused by a virus, is very contagious. There are four million cases of chickenpox a year, with 12,000 hospitalizations and approximately 100 deaths.

Chickenpox itself is not usually serious in children unless they are newborn, are being treated with chemotherapy for cancer, are HIV positive, or are taking steroids. In those cases, the illness can take advantage of the weakened immune system, causing severe illness and even death.

Even though children are usually victim to chickenpox, the virus that causes chickenpox is more serious when it infects adults—especially pregnant women. The symptoms are more severe in adults, the disease lasts longer, and adults are more likely to experience complications. Pregnant women who are close to giving birth can die from chickenpox.

> **Infectious Knowledge**
>
> Chickenpox didn't get its name because of anything having to do with chickens. The red spots people get were thought to look like chick peas. The disease got its name from the Latin word cicer, which means chick peas.

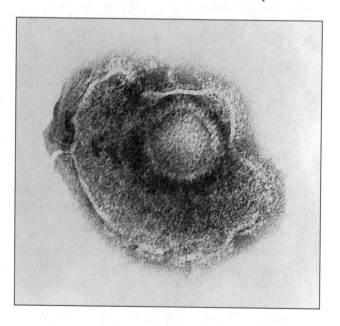

Electron micrograph of a Varicella *(chickenpox)* Virus.

(Courtesy CDC/Dr. Erskine Palmer/B.G. Partin)

Symptoms of chickenpox include an itchy rash that starts in the scalp and spreads to the stomach, back, and face. Victims may run a fever as well. The disease is spread from coughs and runny noses, or direct contact with the fluid from sores of someone who is sick.

After infection, the virus stays in the body for life. Although a person can't get chickenpox twice, the same virus can cause another disease called shingles. Although shingles is most common in adults and people with weakened immune systems, it can occur in younger people as well.

CAUTION

Antigen Alert

Chickenpox is so contagious that an exposed person who isn't immune has a 70 to 80 percent change of getting sick.

Shingles occurs when the virus that causes chickenpox reactivates. Reactivation can occur in the process of normal aging or when the immune system is temporarily or permanently weakened. Symptoms include a painful, blistering rash on one side of the torso or face, pain and numbness, and sometimes tingling two to four days before the rash appears. Pain can last up to a year after the rash is gone. People with weakened immune systems and people over 80 are most susceptible to shingles.

"Mommy, My Ear Hurts"

Earaches are extremely common in young children. Although they are not themselves infectious, they are caused by upper respiratory infections, which are infectious. These infections are spread through contact with respiratory secretions—through coughing and sneezing.

Potent Fact

To help prevent earaches, be sure everyone covers their mouth when they cough, and use a tissue once and throw it away immediately. Keeping toys clean is also a good preventive measure. Try to prevent children from sharing toys that they have put in their mouth, wash toys that children do put in their mouth, and try to teach good hand-washing habits.

The ear aches because of inflammation of the middle ear, caused by fluid that builds up behind the eardrum. Kids often cry, pull on their ears, have a fever, act irritable, and are unable to hear well. In more severe cases, there can be nausea, vomiting, and diarrhea. The developing anatomy of the ear canal in children make some children more susceptible to infection because the ear canal does not drain fluids properly, allowing bacteria to grow. Most children will grow out of this problem.

Sometimes doctors give children antibiotics to cure these earaches, but they don't always work. Many of the organisms that cause the respiratory infections that lead to earaches are antibiotic resistant, and not all respiratory infections are bacterial (remember, antibiotics only help cure bacterial infections). Some children have chronic infections that only respond to surgery to insert a tube that drains fluid from the ear.

Measly Measles

Measles was a very common childhood disease before the introduction of a vaccine in 1963. Today, there are fewer than 1,000 cases per year in the United States. The risk of exposure is much higher in other parts of the world, however.

Early symptoms of measles include fever, runny nose, cough, and sore and red eyes. This is followed by the appearance of a red-brown blotchy rash on the skin. The rash usually starts on the face and spreads downward. It lasts at least three days. Although children who get measles can become fairly sick, most of them recover with no long-term negative effects. Once in a while, measles can lead to pneumonia or encephalitis (swelling of the brain), and permanent disability or death. Adults and very young children are most severely affected.

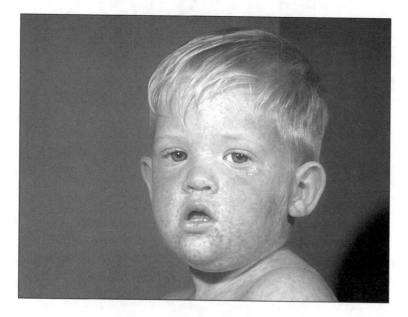

The face of a boy with measles. This is the third day of the rash.

(Courtesy CDC)

Measles, which is caused by a virus, is preventable with a vaccine. It is usually administered as part of the MMR (measles, mumps, rubella) vaccine series, starting when children are 12 to 15 months old, with boosters at 4 to 6 or 11 to 12.

Measles is highly contagious and spreads easily when an infected person coughs or sneezes. Measles particles can stay in the air, so it's possible to get infected by being in a room where an infected person has been.

Down in the Dumps with Mumps

Mumps, which got its name from an Old English word meaning grimace, is also caused by a virus. The incubation period is 12 to 25 days, but 20 to 40 percent of those infected show no symptoms. In 1964, 213,393 cases of mumps occurred in the United States. Today there are between 4,500 and 13,000 cases per year.

When a person infected with mumps does have symptoms, they include a low-grade fever and swelling or tenderness of one or more of the salivary glands in the cheeks and under the jaw. The pain is often worse when swallowing, talking, or chewing. Loss of appetite is also common.

Child with mumps.

(Courtesy CDC/Patricia Smith, Barbara Rice)

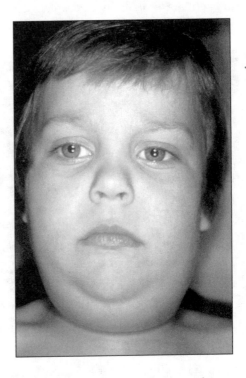

The disease is contagious from three to four days before to four days after symptoms appear. Between 20 and 40 percent of infected people do not have symptoms of mumps. Although mumps doesn't usually cause long-term problems, some of the symptoms, like severe swelling of the salivary glands, can be very uncomfortable. Women may be at risk for spontaneous abortions if they get mumps while pregnant. Mumps is spread from person-to-person through direct contact with saliva, coughs and sneezes, and urine of an infected person.

Fortunately, mumps is preventable with the MMR vaccine, which is discussed later in this chapter.

The "R" in MMR: Rubella

Rubella, also called German measles, is a very contagious disease caused by a virus. The virus causes fever, swollen lymph nodes behind the ears, and a rash that starts on the face and then spreads to the torso and arms and legs. Rubella is not usually serious in children, but can be very serious if a pregnant woman becomes infected. Fortunately, rubella is no longer common, because since 1969 most children are vaccinated for it with the MMR vaccine when they are a year old.

Rubella is spread when a person breathes in droplets of coughs or sneezes from an infected person. A person can spread the disease for several days before the rash appears and for up to a week after the rash goes away.

Rubella is caused by a different virus than the one that causes regular measles, so children need to have both a measles and a rubella vaccine.

Infectious Knowledge
In 1964 and 1965, 20,000 babies were born with birth defects due to a rubella outbreak. The outbreak also caused 10,000 miscarriages and stillbirths.

Potent Fact

The MMR and DPT vaccines each protect against three diseases:

MMR: measles, mumps, rubella

DPT: diphtheria, pertussis, tetanus

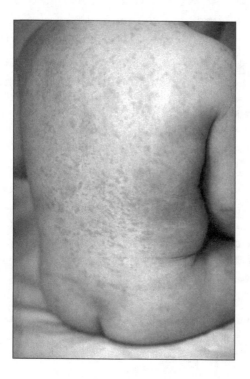

Rash of rubella on the skin of a child's back. Distribution of the rash is similar to that of measles, but the lesions are less intensely red.

(Courtesy CDC)

Deadly Diphtheria

Diphtheria is a bacterial disease that invades the throat. It is spread through contact with salivary or nasal secretions, such as coughs and sneezes, from someone who is infected. The incubation period is seven days, and the major symptom is a sore throat. If the infection moves to the windpipe or respiratory tract, the disease can be much more serious. Complications can include damage to heart muscles and peripheral nerves.

Diphtheria was once one of the most common causes of death in children, but since the introduction and widespread use of the diphtheria vaccine, it is rarely found in the United States. However, some children in the United States aren't properly vaccinated, so some cases do occur.

The disease is still common in some parts of the world, including the Caribbean and Latin America. During the last few years, large epidemics of diphtheria have occurred in the former Soviet republics, as well as Algeria, China, and Ecuador. The majority of cases in many of these epidemics have been in young people who were improperly vaccinated, or not vaccinated at all.

In the United States, the diphtheria threat is shifting from children to adults and adolescents. Cases are occurring in people who have not been immunized and in vaccinated people who did not receive periodic booster doses to maintain their immunity. Routine vaccination of both children and adults is essential to prevent the reemergence of diphtheria in the United States.

Pertussis, or Whooping Cough

Pertussis, or whooping cough, was a major cause of illness and death among infants and children in the United States before vaccines were introduced in the 1940s. Following the introduction and widespread use of the combined pertussis, diphtheria, and tetanus vaccine (DTP) among infants and children in the late 1940s, the incidence of reported pertussis declined to a historic low of 1,010 cases in 1976.

Disease Diction

Whooping cough got its name from the whooping sound children make when they try to breathe after a coughing spell.

Whooping cough is a very contagious and dangerous respiratory infection caused by the *Bordetella pertussis* bacterium. Symptoms of whooping cough generally include runny nose and a cough that gets worse and worse. Violent coughing spells can end with vomiting. Once the whooping stage begins, antibiotics don't work.

Whooping cough is spread through the air, making it particularly infectious.

Terrifying Tetanus

Tetanus, also called lockjaw, is very rare in the United States because most children are vaccinated against it. Tetanus is caused by a bacterium that is common in the soil, but dies quickly when it is exposed to oxygen. People who haven't been vaccinated for tetanus can get the disease by stepping on a dirty nail or getting cut by a dirty tool. The bacterium produces a toxin, or poison, that spreads in the bloodstream and can result in severe muscle spasms, paralysis, and death.

Tetanus is difficult to treat, but proper vaccination prevents it. Children get a tetanus shot in combination with pertussis and diphtheria vaccines. Adults need a booster shot every ten years to make sure they are protected.

Potent Fact

If you have an open cut, you should figure out when you had your last tetanus booster. If it is more than 10 years ago, it's time to get vaccinated again.

Infectious Knowledge

According to the Centers for Disease Control & Prevention (CDC), the following is the recommended childhood immunization schedule.

- **Hepatitis B:** Four doses total. The first one at birth, the second at least a month later, the third 12 weeks after that, and the final dose after the baby is at least six months old. (See Chapter 11 for more on Hepatitis B.)
- **Diphtheria, pertussis, tetanus (DPT):** Starting at two months, again at four months, and a third dose at six months. The fourth dose must be at least six months after the third dose. A booster is recommended at age 11, with a routine booster every ten years afterward.
- **Flu:** The first flu shot can be as early as age two months. After that, annual vaccination should occur. A pneumococcal vaccine that prevents 60 to 70 percent of bacterial pneumonia is given with the flu vaccine when the child is two.
- **Polio:** Four doses at two months, four months, six to eighteen months, and four to six years.
- **Measles, mumps, rubella (MMR):** The first dose is at one year. The second dose is recommended at four to six years.
- **Chickenpox:** Vaccine at one year.
- **Hepatitis A:** Recommended for certain states and regions and for high-risk groups. Ask your doctor whether your child needs this vaccine. (See Chapter 11 for more on this disease.)

The Importance of Childhood Vaccination

With the advent of a variety of effective childhood vaccinations, we now have very low rates of many of once deadly childhood diseases. However, they haven't completely disappeared. The bacteria and viruses that cause them are still around—in the United States and elsewhere in the world—so it's extremely important that children are properly vaccinated.

A number of these vaccines are given in combination. As noted earlier in this chapter, MMR is a vaccine against measles, mumps, and rubella. DPT is a vaccine against diphtheria, pertussis (whooping cough), and tetanus. Flu, hepatitis A and B, polio, and chickenpox vaccines are also available to prevent against those diseases.

The Common Cold

The cold is caused by a wide variety of viruses. Symptoms include sore throat, runny nose and watery eyes, sneezing, chills, and mild aches and pains. Colds are spread when a healthy person breathes in germs that a sick person has coughed, sneezed, or breathed into the air. It's also possible to get a cold if you touch a surface that a sick person has coughed or sneezed on and then touch your eyes or nose.

Potent Fact

There are an estimated one billion colds in the United States each year.

Children are very susceptible to colds. In fact, children are the primary sources and spreaders of cold viruses! Why? Because they are in close contact with other children and adults, they haven't learned the lessons of good hygiene yet so their hands are often not properly washed, and they don't always cover their nose and mouth when they cough or sneeze. In addition, their sinus and ear drainage passages and bronchial tubes are small, so they are easily blocked by mucus and swelling.

Many types of viruses cause colds, but the common cold is usually caused by a family of viruses called rhinoviruses. Each virus may have a slightly different pattern of symptoms and severity. Cold viruses enter the body through the nose. They infect a small number of cells in the lining of the nose, and that is enough to cause infection. The incubation period for a cold is 10 to 12 hours. Peak symptoms usually appear between two and three days later. Colds can last up to two weeks, but if they hang on longer, they may be allergy-related or they may have led to a secondary respiratory infection.

The symptoms that develop after infection with a cold virus are due to the immune system's response to the infection. That response causes secretions from the nose and triggers cough and sneeze reflexes. It also stimulates pain nerve fibers, causing aches and soreness in muscles and joints.

Symptoms of the common cold are obstructed breathing, swelling of sinus membranes, sneezing, coughing, sore throat, and sometimes a low-grade fever.

After years of trying, we still have no cure for the common cold. The best we can do is treat the symptoms. Bed rest, drinking lots of fluids, gargling with warm salt water, and taking aspirin, acetaminophen, or ibuprofen for relief of aches and fever can help ease the discomfort a cold brings. Sometimes anti-histamines, decongestants, and cough medicine help, too. None of these things will shorten the duration of a cold, but they will make it easier to bear.

> **CAUTION**
>
> **Antigen Alert**
> There is no conclusive medical evidence that Vitamin C helps to prevent colds.

Kiss and Tell: Infectious Mononucleosis

Infectious mononucleosis or "mono" is an illness that afflicts teenagers and young adults, mainly ages 14 to 30. It has been estimated that approximately 50 percent of students have had mono by the time they enroll in college. Mono is caused by the Epstein-Barr virus, a member of the Herpes family of viruses, but there are other viruses that may produce a mono-like illness. The disease usually occurs sporadically and outbreaks are rare. Many times the symptoms are so mild it isn't even recognized.

In underdeveloped countries, people are exposed to the virus in early childhood, when they aren't likely to develop noticeable symptoms. In developed countries such as the United States, the age of first exposure may be delayed until older childhood and young adulthood, when symptoms are more likely to occur.

The most common symptoms include excessive fatigue, headache, loss of appetite, sore throat, swelling of the tonsils, enlarged lymph nodes (swollen glands) in the neck, underarms, and groin. A low-grade fever occurs at first, and then rises to above 100 degrees after the third or fourth day. Sometimes, the liver and spleen are affected and enlarged. The disease lasts one to several weeks. A small proportion of affected people can take months to return to their normal energy level.

> **Infectious Knowledge**
>
> The Epstein-Barr virus is found in moist exhaled air and secretions from the nose and throat. It isn't as contagious as many other viruses but may be transmitted by direct contact, which explains the origin of mono's nickname as the "kissing disease."

How Soon Do the Symptoms Appear?

Symptoms appear from four to six weeks after exposure, but may be the same as many other illnesses, such as the common colds or strep throat. For this reason, it is particularly difficult to diagnose mono in the early stages of illness. The diagnosis is helped by two blood tests: one that looks at an increase in a specific type of white blood cell and another that identifies an antibody which is present when a person has mononucleosis.

No treatment other than rest is needed in the vast majority of affected people. Due to the risk of rupture of the spleen, contact sports should be avoided until clearance has been given by a physician. On rare occasions, a short-term course of steroids like prednisone may be of value for extreme throat swelling that inhibits swallowing or endangers breathing. Steroids don't cure the disease, but serve to reduce the inflammatory response. In most cases of mononucleosis, hospitalization isn't necessary.

Currently, there is no vaccine available to prevent infectious mononucleosis. People who have had mono can shed the virus periodically in their saliva for the rest of their lives.

The Least You Need to Know

- Many childhood diseases that used to be deadly can now be prevented with proper vaccinations.
- Some infections, like diphtheria, are reemerging as problems in other parts of the world where vaccination programs are not as strong as they are in the United States.
- It's crucial that all children are properly vaccinated.
- Children are the most common carriers and spreaders of the common cold.
- There is no cure for the common cold, only treatment of symptoms.

Chapter 16

Vacationers' Nightmares: Tropical Diseases

In This Chapter

- Where to watch for leishmaniasis
- Avoiding giardiasis
- The dangers of schistosomiasis
- What causes river blindness

Many diseases are caused by organisms that thrive in tropical climates or live inside flies that are common in the tropics. Many of these so-called "tropical diseases" are treatable; however, developing countries don't always have access to the medications they need to treat them. Consequently, many people unnecessarily suffer and even die from tropical diseases.

In this chapter, you will find out about some of the most common tropical diseases, their treatment, and how to prevent them.

Leishmaniasis: The Sand Fly's Bug

Leishmaniasis is a parasitic disease caused by a protozoan that initially lives in the sand fly and is transmitted to people through sand fly bites. The organism develops and multiplies in the gut of the fly and is introduced into the bloodstream of humans after a bite. It can cause a skin infection or a more serious systemic infection. The skin infection, which consists of sores, develops weeks or months after a sand fly bite. The more serious infection, which consists of fever, enlargement of the liver and spleen, and anemia, can take months or even years to develop.

Potent Fact

Leishmaniasis affects between 12 and 15 million people, with one million new cases reported annually.

The disease is found in 90 tropical and subtropical countries around the world. More than 90 percent of the systemic cases occur in Bangladesh, Brazil, India, Nepal, and Sudan. The disease is rare, but not unheard of, in the United States.

Skin Leishmaniasis

The incubation period after a sand fly bite ranges from a couple of weeks to several months, with no symptoms to indicate the presence of the disease. A cut or some other trauma to the skin can result in activation of a skin infection a long time after the initial bite. The skin lesions, which can be painful, usually evolve over time. They are usually found on exposed areas of the skin, especially the arms, legs, and face. When the lesions heal, they leave scars. The healing process can take months.

Systemic Leishmaniasis: The Black Fever

Malnutrition has been shown to contribute to this more serious form of the disease. The fever that develops may be continuous, or it may come and go at irregular intervals. Symptoms include weight loss, diarrhea, and weakness. Darkening of the skin is also a characteristic of the disease, which explains why it is sometimes called "black fever."

The disease can be diagnosed in a variety of ways, including a skin test, which is simple and sensitive, but can't distinguish between active, inactive, new, or old infections. The disease-causing organism can be identified by examining a stool sample under a microscope, and cultures can be grown and antibody tests can be done as well.

CAUTION

Antigen Alert

People with HIV/AIDS are more susceptible to leishmaniasis.

The skin lesions caused in the mild form of the disease usually heal by themselves in 2 to 10 months; however, they may leave scars. Good wound care (use of local heat and keeping wounds clean) is important to reduce the chance of permanent scarring.

Antiparasitic drugs can be given as therapy once someone is infected, but they don't prevent the disease. Unfortunately, these drugs are not readily available in the developing countries where the disease is most common. Antibiotics and antifungal drugs have been used to treat the disease with some success as well.

Risk for Travelers and Prevention

Anyone who lives in or travels to places where the organism is found is at risk for contracting leishmaniasis. The highest risk is to those who are outside between dusk and dawn, which is when the sand fly is most active.

There are no vaccines or drugs that prevent the disease; the most effective preventive measures are to reduce contact with sand flies. Although sand flies mostly bite at night, they will bite during the day if they are disturbed. Travelers should wear protective clothing and use insect repellent. Bed nets and screen doors and windows should be used as well. The netting must be very fine in order to be effective, as sand flies are about one third the size of mosquitoes.

After treatment for infection, approximately 98 percent of people have immunity against reinfection with the same strain.

Disfiguring Complications

Complications for leishmaniasis can include secondary bacterial infection; disfiguring of the nose, lips, and palate; bleeding, ruptured spleen; edema (swelling); and lesions in the nose and throat, with destruction of tissue.

The prognosis for those who get leishmaniasis depends on their immune status as well as whether they are getting proper nutrition. People with weakened immune systems or who suffer from malnutrition are at greater risk of severe leishmaniasis or developing complications. With early treatment, 90 percent of people are cured. The death rate is 15 to 25 percent in untreated cases, with death occurring in 3 to 20 months.

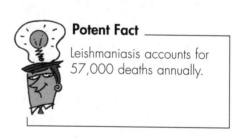

Potent Fact

Leishmaniasis accounts for 57,000 deaths annually.

Giardiasis: A One-Celled Wonder

Giardia is a one-celled, microscopic parasite that lives in the intestines of people and animals and is passed in the stool of an infected person or animal. It is protected by an outer shell, which helps it survive outside the body and in the environment for long periods of time. During the past 20 years, giardia has been recognized as one of the most common causes of waterborne disease in the United States.

The giardia parasite causes a disease called giardiasis, which results in diarrhea, belching, gas, and cramps. The disease is easy to catch if you drink untreated spring water or water from a stream. Many animals carry giardia in their feces and may introduce it into rivers, streams, and springs. Infected water may look clean and safe even if it isn't. City water supplies can get infected if sewer lines rust or leak. It is also possible to get the disease from drinking water in certain countries if that water hasn't been boiled or treated.

Symptoms usually develop one to two weeks after infection, and they last two to six weeks.

How Does It Spread?

Giardia lives in the intestines of infected people and animals. Millions of germs can be excreted in a single bowel movement from a person or animal. The parasite can be found in soil, food, water, or surfaces that have been contaminated with feces from infected people or animals. Infection can happen after accidentally swallowing the parasite.

Giardia is spread by …

- Putting something in your mouth or accidentally swallowing something that has come in contact with the stool of a person or animal that is infected.
- Swallowing water that is contaminated. This could be water from a swimming pool, hot tub, Jacuzzi, fountain, lake, river, spring, pond, or stream that can be contaminated with sewage or feces from humans or animals.
- Eating uncooked food contaminated with giardia.
- Accidentally swallowing giardia picked up from surfaces contaminated with stool from an infected person. Contaminated surfaces could be toys, bathroom fixtures, changing tables, or diaper pails.

Giardiasis is diagnosed by examining a stool sample under a microscope and looking for the presence of the organism. Sometimes several samples have to be taken before the diagnosis can be made.

The disease is usually treated with a medicine called Flagyl, which is typically taken three times a day for 5 to 10 days. Side effects include nausea and a metallic taste in the mouth. You shouldn't drink alcohol while taking the medicine, and it shouldn't be taken in the early stages of pregnancy. Young children are sometimes given a different medicine.

Antigen Alert

Everyone is at risk for giardiasis, which is a highly contagious disease.

Because the disease is so contagious, sometimes whole families are treated together even if they don't all show symptoms for the disease. The doctor will probably check another stool sample when treatment is finished in order to be sure that the medicine worked. There are situations where a second medication or a longer course of treatment may be necessary.

Potent Fact _____

According to the CDC, if you are diagnosed with a giardia infection, you can prevent it from spreading through several simple steps:

♦ Wash your hands with soap and water after using the bathroom, changing diapers, and before eating or preparing food.

♦ Avoid swimming in recreational water for at least two weeks after diarrhea stops (you can still pass the infection after your symptoms have stopped).

♦ Avoid fecal exposure during sex.

Wash Those Hands!

Practicing good hygiene is the best way to prevent infection. Wash your hands with soap and water after using the bathroom and before handling or eating food. Wash your hands after every diaper change, especially if you work with young children. Protect others by not swimming if you have diarrhea and not letting your children swim if they have diarrhea.

Another good preventive measure is to avoid water that might be contaminated. Don't swallow water when you are swimming. Don't drink untreated water from wells, lakes, rivers, springs, or ponds. If there is an outbreak in your community, don't drink untreated water. Avoid using ice or drinking untreated water when traveling in countries where the water supply may be unsafe. If you have to drink untreated water, treat it yourself by heating the water to a rolling boil for at least one minute. Avoid food that might be contaminated. Wash and/or peel all raw fruits and vegetables before eating them. Avoid eating uncooked foods in countries with minimal water treatment and sanitation systems.

Schistosomiasis: Snail Fever

Among human diseases caused by parasites, schistosomiasis ranks second behind malaria in terms of its social, economic, and public health impact in tropical and subtropical regions of the world. One hundred twenty million people have symptoms, and 20 million suffer severe consequences from the disease.

Schistosomiasis was first recognized in the time of the Egyptian pharaohs. The worms that cause the disease were discovered in 1851 in a hospital in Cairo by Theodor Bilharz, a German pathologist. The disease was originally named bilharziasis after him.

Schistosomiasis is endemic in 76 tropical developing countries, and 600 million people are at risk for acquiring the disease. There are estimates that up to 200 million are already infected. Extreme poverty and poor sanitary conditions are major risk factors for the disease, along with inadequate public health infrastructure.

The disease affects many children between the ages of 10 and 19. It also affects farmers and freshwater fishermen. Water resource development can also aggravate the disease.

> **Infectious Knowledge**
>
> Between 1950 and 1990, the number of dams in the world increased from 5,000 to more than 36,000. There has been a corresponding increase in schistosomiasis, especially in sub-Saharan Africa.

The major forms of schistosomiasis are caused by five different species of waterborne flatworms called schistosomes. They cause several variations of the disease.

The worms enter the body through contact with infested water. This can be by washing hands, washing food, swimming, fishing, farming, and growing rice. Migration is introducing the disease into more urban areas in northeast Brazil and Africa, and refugees moving around are spreading it in Somalia and Cambodia. Tourists are also at risk for the disease.

Signs of Schistosomiasis

The urinary form of the disease is characterized by the presence of blood in the urine, which can lead to bladder cancer or kidney problems. The intestinal form of the disease is characterized by intermittent, bloody diarrhea, which can lead to serious complications of the liver and spleen. Those who get the disease are very weakened by it and are often unable to work.

Within days of infection, people develop a rash or itchy skin. Fever, chills, cough, and muscle aches can start within a month or two of infection. Most people have no symptoms during the early stages of infection.

Infected individuals can contaminate their environment. The worm eggs in human excrement hatch on contact with water and release larvae. The tiny larvae must find a freshwater snail to survive. Once inside the snail, the larvae divide several times and produces thousands of new parasites. The new parasites are excreted by the snails into the surrounding water. It only takes a few seconds for these worms to penetrate human skin.

> **Potent Fact**
>
> Female long worms can lay 200 to 2,000 eggs per day over a five-year period!

From the skin, the parasites get into the bloodstream. Within 30 to 45 days, they grow into long worms (12 to 16 millimeters in length). In intestinal disease, the worms live in the blood vessels that line the intestine. In urinary disease, they live in the blood vessels of the bladder. Only half of the eggs are excreted in feces or urine. The rest are trapped in the body tissues and do damage to vital organs. It is the eggs, not the worm, that damage the intestines, bladder, and other organs.

The disease is diagnosed by analysis of a stool or urine sample to see if the parasite is present. A blood test is also available. It takes six to eight weeks after exposure to contaminated water for the blood test results to be accurate.

Fast and Easy Treatment

There are now three safe and effective drugs to treat schistosomiasis. Patients have to take pills for only one or two days.

Don't Drink the Water

There are a number of ways to prevent the disease. According to the CDC, they include the following:

- Avoid swimming or wading in fresh water when traveling in countries where the disease is common.

- Drink safe water.

- Bath water should be heated for five minutes at 150 degrees. Water held in a storage tank for at least 48 hours should be safe for showering.

- Vigorous towel-drying after an accidental, brief water exposure may help prevent the parasite from penetrating the skin, but this method should not be relied upon to prevent exposure and disease.

> **Infectious Knowledge**
>
> Schistosomiasis is endemic in the following areas: Southern Africa, sub-Saharan Africa, Lake Malawi, the Nile River Valley, Brazil, Suriname, Venezuela, Antigua, Dominican Republic, Guadeloupe, Martinique, Montserrat, Saint Lucia, Iran, Iraq, Saudi Arabia, Syria, Yemen, Southern China, Philippines, Laos, Cambodia, Japan, Central Indonesia, and the Mekong Delta.

African Sleeping Sickness

Human African trypanosomiasis, known as sleeping sickness, is caused by trypanosomes, which are protozoan parasites spread by the tsetse fly. The flies live in Africa and are found in vegetation by rivers and lakes, forests, and wooded savannah.

Sleeping sickness is a daily threat to more than 60 million men, women, and children in 36 countries of sub-Saharan Africa, 22 of which are among the least developed countries in the world. Some 300,000 to 500,000 people may develop disease each year.

Another human form of the disease, human American trypanosomiasis, occurs in the Americas and is known as Chagas disease.

When a person becomes infected they have bouts of fever, headaches, pains in the joints, and itching as the parasites multiply in the blood and lymph glands. In the second phase of the disease, the parasite crosses the blood-brain barrier and infects the central nervous system. This is when the characteristic signs and symptoms of the disease appear including confusion, sensory disruption, and poor coordination. Disturbance of the sleep cycle, which gives the disease its name, is the most salient feature.

Without treatment, the disease is usually fatal. If the disease is diagnosed early, the chances of cure are high. However, if the patient does not receive treatment before the onset of the second phase, neurological damage is irreversible even after treatment. Unfortunately, treatment of trypanosomes is leading to increased drug resistance, which limits therapeutic success.

River Blindness

River blindness is caused by a parasitic worm that lives for up to 14 years inside the human body. This disease is called river blindness because the black fly vector lives in fertile areas near rivers. Each adult female worm produces millions of larvae that migrate throughout the body and cause serious visual impairment and sometimes blindness, rashes, lesions, intense itching, depigmentation of the skin, elephantiasis, and general debilitation. The disease shows symptoms one to three years after the infectious larvae enter the body.

> **Infectious Knowledge**
>
> In 1987, Merck & Co., Inc., the manufacturer of ivermectin, pledged to provide the drug free-of-charge for as long as necessary to overcome river blindness as a public health problem. It established a donation program that works with the World Health Organization, health ministries, and nongovernmental organizations. Between 1987 and the end of 1996, more than 65 million doses of the drug had been donated for distribution.

The parasite is carried by the black fly, which lays its eggs in the water of fast-flowing rivers. Adults emerge after 8 to 12 days and live for up to four weeks. The female black fly ingests the larvae if she bites an infected person. She can spread the parasite when she bites other people afterward.

A total of 18 million people are infected; 99 percent of them are in Africa. Of those infected, more than 6.5 million suffer from severe itching or dermatitis and 270,000 are blind.

Ivermectin to the Rescue

The development of a drug called ivermectin in the 1980s was the first time there was a safe, effective drug that could improve symptoms and decrease the chances of disease transmission.

It turns out that a single dose of ivermectin once a year is all that is needed!

The Least You Need to Know

◆ Travelers should wear protective clothing and use insect repellent to prevent leishmaniasis, which is caused by sand fly bites.

◆ Giardiasis can be prevented by frequent hand washing, especially after using the bathroom or changing a diaper, and refraining from using a swimming pool while infected.

◆ Schistosomiasis ranks second behind malaria in terms of its social, economic and public health impact in tropical and subtropical regions of the world but can be easily treated with medication once diagnosed.

◆ Trypanosomiasis, known as sleeping sickness, is a daily threat to more than 60 million men, women, and children in 36 countries of sub-Saharan Africa.

◆ Black flies carry and transmit the parasite for river blindness, which is primarily found in Africa.

Chapter **17**

Meals You'll Never Forget: Food-Borne Diseases

In This Chapter

◆ The five most common causes of food-borne illness

◆ How infections are spread

◆ Rare but debilitating side effects of food-borne infections

◆ Simple steps to take to prevent infection

In 1993 in Seattle, Washington, several school-aged children became ill, suffering from diarrhea and stomach cramps. This was no flu outbreak, though. Epidemiologists determined that the cause of the illness was a bacterium called *E. coli* 0157:H7. The children had all eaten infected hamburgers at the same Jack in the Box restaurant. Overall, 500 people in the Pacific Northwest got *E. coli* 0157:H7 infections that year, and three of the children from the Seattle outbreak died.

The CDC used DNA fingerprinting to track the bacteria and link the illnesses in people to undercooked hamburger patties from a Jack in the Box restaurant. Hamburgers from the restaurant were recalled, preventing further illness.

The Jack in the Box incident got a lot of media attention because of the number of children who got sick and because the source of the infection was a large chain restaurant. However, such food-borne illnesses are common throughout the world, and the incidents rarely ever get media attention.

Many different bacteria—all of them with complicated-sounding names—are responsible for causing food-borne diseases. In this chapter you'll read about six of the most common bacteria. Fortunately, most food-borne illnesses don't last long and aren't very dangerous, but there are cases where there are possible serious long-term effects, which we'll address, too.

E. coli 0157:H7

The *E. coli* bacterium lives everywhere in the environment, though it is particularly common in animals. In people, *E. coli* live in parts of our body that are exposed to the environment, such as our intestines and respiratory tract. In our intestines, the bacteria often live at peace with us and in fact help us by being sources of vitamins K and B complex.

There are hundreds of strains of *E. coli*. They are classified and numbered by the antigens that they produce. Some of them have acquired genes, either through mutation or from other organisms, that allow them to produce chemicals that are harmful to people. If we ingest these harmful strains, like 0157:H7, they can make us sick by producing these powerful chemical poisons, or toxins. However, most *E. coli* strains are harmless and live in the intestines of healthy humans and animals.

When people eat *E. coli*-infected food or come directly into contact with *E. coli*-infected fecal matter, the *E. coli* bacteria enter the body and make their way to the stomach and small intestine, and often attach to the inside surface of the large intestine. Toxins, or poisons the bacteria secrete, cause swelling of the intestinal wall, which is what causes severe gastrointestinal distress.

> **Infectious Knowledge**
>
> *E. coli* is often used by researchers as a basic research tool because it grows quickly. Researchers study its functions and how it reproduces to learn more about bacteria in general, and it is used as a model organism because its behavior is similar to other disease-causing bacteria.

Painful and Bloody: Hemorrhagic Colitis

E. coli 0157:H7 causes a disease called hemorrhagic colitis, which is the sudden onset of stomach pain and severe cramps. This is followed by diarrhea that is watery and bloody. Sometimes there is vomiting, but there is no fever. The incubation period is three to nine days. The illness lasts about a week, and there are usually no long-term problems.

Antibiotics have little if any effect on this disease and most people recover in 5 to 10 days. There is no specific therapy, but it's important to drink lots of water to stay hydrated, and to eat properly.

More Severe Cases

A small percentage of people with *E. coli* infections get a more serious condition, called hemolytic uremic syndrome (HUS), that can be life-threatening. It develops when the bacteria gets into the circulatory system through the inflamed bowel and releases certain toxins into the blood.

HUS takes one to two weeks to develop, and 50 percent of the people in the United States who come down with the illness die from it. Half of all people who get HUS need dialysis for the rest of their lives, and many infected individuals need blood transfusions.

E. coli is diagnosed through laboratory analysis of a stool sample.

> **Potent Fact**
> HUS is the most common cause of kidney failure in children.

Where Does It Come From?

E. coli lives in the intestines of cattle, chicken, deer, sheep, and pigs. Animals are just carriers—*E. coli* doesn't make them sick. The use of untreated animal manure as fertilizer is a common route of transmission for the bacterium.

E. coli can be spread by eating ground beef, unpasteurized juice or milk, alfalfa sprouts, or water. Person-to-person transmission can occur in places like day care centers, hospitals, and nursing homes, or anywhere people come into contact with fecal matter of an infected individual.

Unlike many infectious organisms, where it takes thousands or tens of thousands of organisms to cause disease, it only takes a few organisms, fewer than 200, for an *E. coli* infection to occur.

> **Antigen Alert**
> Contaminated meat looks and smells like normal meat, so thorough cooking of food is necessary to prevent disease. Ground meat is more of a risk than whole cuts because in the former the bacteria are mixed in the grinding process and may not be completely killed by cooking, whereas in the latter they are located only on the surface and are more easily killed.

Keeping Infection at Bay

The biggest risk for *E. coli* infection is eating undercooked beef. Beef should always be thoroughly cooked before eating. Other ways to prevent the ingestion of *E. coli* include ...

- Avoiding spreading harmful bacteria in the kitchen by keeping raw meat separate from other food. Wash hands, counters, cutting boards, and utensils with hot, soapy water after they touch raw meat. Never put cooked meat on the same unwashed plate that held raw meat (this is important for people who grill burgers and put the grilled burgers on the same platter they brought the raw burgers out on).

- Washing meat thermometers after each and every use.

- Drinking only pasteurized milk and cider.

- Washing fruits and vegetables thoroughly.

- Drinking only purified, treated water.

It's a Mouthful: *Camphylobacter*

Camphylobacter is the most common cause of food-borne illness. The bacterium was first identified as a cause of food-borne illness in 1975, and is most commonly spread by poultry.

Potent Fact

Camphylobacter and other food-borne diseases are often undiagnosed and underreported, so it is difficult to estimate the exact numbers of people who get sick each year.

Most cases are isolated or part of small outbreaks. Seventy percent of cases are the result of eating under-cooked chicken, although unpasteurized milk, under-cooked meat, mushrooms, hamburger, cheese, port, shellfish, and eggs can also cause illness.

Camphylobacter normally lives in the intestines of mammals and warm-blooded birds. It can survive refrigeration and grows if food is left out for too long at room temperature. The organism is sensitive to heat, so proper cooking and pasteurization will kill it.

Diarrhea, Again

Diarrhea is the most common symptom of infection. Other symptoms include fever, nausea, vomiting, stomach pain, headache, and muscle pain. The incubation period is two to five days. The illness usually lasts a week, but can stay around for up to three weeks.

If the person who is sick has AIDS, *Camphylobacter* infection can be quite severe.

Like *E. coli*, *Camphylobacter* infection is diagnosed by laboratory analysis of a stool sample.

Antibiotics are effective in treating *Camphylobacter* infections, and antidiarrheal medication is recommended as well. It's also important to drink lots of fluids to prevent dehydration.

The Immune System's Terrible Mistake

Some people who get *Camphylobacter* infections develop a rare and paralyzing nerve disease called Guillain-Barré syndrome. Guillain-Barré is the result of a terrible mistake made by the immune system. The antibodies our body makes to fight the infection sometimes attack the body's own nerve cells, because the nerve cells are chemically similar to the disease-causing bacteria and our immune system can't always distinguish between the two. The damage to the attacked nerve cells causes paralysis.

The paralysis starts in the feet and spreads up the body. Sometimes full paralysis occurs and lasts for months. Often patients must be hospitalized in intensive care units for long periods of time. Full recovery is common, but some people are left with severe, permanent nerve damage. Fifteen percent of people with Guillain-Barré have paralysis that causes them to remain bedridden or in wheelchairs a year after coming down with the syndrome.

One in 1,000 people who get *Camphylobacter* infections get Guillain-Barré.

> **Antigen Alert**
>
> Another possible long-term side effect of *Camphylobacter*, *Salmonella*, *Shigella* (discussed later in this chapter), and other food-borne bacterial infections is called Reiter's syndrome. It's a form of arthritis that primarily affects the knees and lower back. It can become a chronic condition.

Cook That Chicken!

Infection control at all stages of food processing is important, but in order to be sure to prevent infection, it's important to cook poultry properly.

Chicken should be put in the coolest part of the car on the way home from the grocery store, it should be defrosted in the refrigerator or microwave (not left out on the counter), stuffing should be cooked outside the bird, and leftovers should be cooled quickly.

Cooked chicken shouldn't be left out at room temperature for more than two hours. The same rules for protecting against *E. coli*, such as hand-washing, drinking only pasteurized milk and treated water, and washing fruits and vegetables, are true for this organism as well.

> **Potent Fact**
>
> The United States Food and Drug Administration recommends using a cooking thermometer to be sure the thickest piece of meat is at least 180°F.

> **Potent Fact**
>
> When bagging your groceries, bag frozen foods and meats together. It's a safer way to transport meats.

Salmonella (No, It's Not Named After the Fish)

The *Salmonella* bacterium was first isolated from a pig's intestine by American veterinarian Dr. Daniel Salmon. It's a common bacterium that causes a variety of intestinal infections, including typhoid fever, although the strain that causes typhoid is very rare in the United States. There are many varieties of *Salmonella*, and some are now becoming resistant to the antibiotics we use to treat them. Salmonella can be found in raw eggs and a variety of foods. Like other food-borne organisms, it gets into the body when we eat.

Disease Diction

An **enteric infection** is one that affects the intestines.

Salmonella is one of the most common causes of *enteric*, or intestinal, infections. There are 40,000 cases diagnosed each year, and many mild cases probably go unnoticed.

Salmonella's Symptoms

The incubation period for a *Salmonella* infection is 6 hours to 10 days. Symptoms usually show up in 6 to 48 hours. They include diarrhea, fever, and stomach cramps. Sometimes they start with nausea and vomiting.

Babies, the elderly, and those with compromised immune systems are most susceptible to infection.

Salmonella is rarely fatal, and like the other food-borne diseases we've discussed in this chapter, it's diagnosed with laboratory analysis of a stool sample.

Antibiotics are not recommended unless the infection has spread from the intestines, because such medication can prolong rather than reduce the period of bacterial shedding in the intestine. Individuals usually feel better within five to seven days.

Be a Clean Chef

Proper handling of eggs and other foods is key to preventing *Salmonella* infection. Since the organism spreads rapidly, it's also important to report cases quickly. You should be sure all foods are kept separate to prevent cross-contamination. All foods should be cooked thoroughly. Raw eggs should not be eaten.

As with the other food-borne organisms, drink pasteurized milk and purified water and be sure to wash hands, surfaces, utensils, and cutting boards after every use.

Deadly *Shigella*

Shigella is a bacterium that thrives in the intestines and causes sudden, severe diarrhea. It was discovered more than 100 years ago by a Japanese scientist named Kryoshi Shiga.

Shigella is spread in foods like salads, raw vegetables, milk and other dairy products, and poultry; but it can also be spread from person to person. There are 25,000 reported cases in the United States every year, but it is probably seriously underreported. Some estimates are as high as 450,000 cases per year.

Children and those with HIV infections are more susceptible to *Shigella* infection. The disease usually comes and goes fairly quickly, but antibiotics help to cure it more quickly.

You Guessed It: More Diarrhea

The symptoms of a *Shigella* infection are fever, stomach cramps, and diarrhea that is bloody and has mucus in it.

The incubation period is 12 hours to six days, but people usually get sick in one to two days.

Shigella multiplies in the gut, invades cells, and causes tissue destruction. It's more severe than most other food-borne illnesses, because the poisons the bacteria secrete can do severe damage.

The disease usually goes away in about a week, but it can be months before someone who was sick feels normal again.

Some cases of *Shigella* require hospitalization. These more serious cases have symptoms like dehydration, seizures, rectal bleeding, and bacteria in the bloodstream.

> **Infectious Knowledge**
>
> One million people a year die from *Shigella* infection worldwide. The majority of them are children in developing countries.

Source of Infection

Shigella is passed from the infected stool of a previously infected person. It's spread through person-to-person contact (80 percent of infections) or in contaminated food or water (20 percent of the time).

Like *E. coli*, an infection can occur with only a small number of bacteria entering the body.

Like the other food-borne diseases, *Shigella* is diagnosed by laboratory analysis of a stool sample.

It's sometimes hard to isolate *Shigella* bacteria because they are similar to other bacteria that normally live in our colons.

Shaking Down *Shigella*

To prevent *Shigella* infection …

◆ Always wash hands after going to the bathroom and before eating. Adults need to wash their own hands carefully and supervise children so they wash their hands properly as well.

◆ When changing diapers, be sure to dispose of them properly, wash hands with warm, soapy water, and clean the surface where the baby was changed.

◆ Maintain proper levels of chlorine in pools.

◆ Practice basic food safety.

◆ When traveling in the developing world, drink only bottled water, avoid ice, and eat only food that has been cooked.

Shigellosis can usually be treated with antibiotics such as ampicillin and ciprofloxacin. However, many *Shigella* strains are resistant to antibiotics, which limits the effectiveness of therapy. Using antibiotics to treat shigellosis without proper evaluation of drug susceptibility will further exacerbate the resistance problem.

Looking for *Listeria*

Doctors and epidemiologists often overlook *Listeria*, another food-borne bacterium, as a cause of food-borne illness. This is because the bacterium is difficult to grow in the laboratory, and it is often confused with harmless contaminants that can grow in a culture and therefore are disregarded. *Listeria* can survive at low temperatures. In other words, it can be transmitted in ready-to-eat foods even if they have been properly refrigerated.

Listeria live everywhere in the environment. They are in grazing areas, stale water supplies, and poorly prepared animal feed. They also live in the intestines of people, animals, and birds without causing disease. They've been found in cattle, sheep, fowl, dairy products, fruits, and vegetables.

Infectious Knowledge
There are 2,500 cases of *Listeria* in the United States each year, which result in 500 deaths.

Listeria infection is caused by eating contaminated food. The bacterium is often opportunistic, living peacefully in the body until another disease weakens the body's immune system, when it takes advantage of the situation and causes disease. Pregnant women are also more at risk for *Listeria* infection.

The bacteria travel through the bloodstream and are often found inside of cells. They are able to use the machinery of our cells to avoid our immune system's efforts to fight them. Then they produce toxins, or poisons, that damage our body's cells.

Aches, Nausea ... and More Diarrhea

The incubation period for a *Listeria* infection is one to eight weeks. Most people start to see symptoms a month after initial infection.

Symptoms include fever, muscle aches, nausea, and diarrhea. If the organisms get into the nervous system, symptoms include headache, stiff neck, loss of balance, confusion, and convulsions. If the organisms get into the brain, symptoms mimic those of a stroke.

For pregnant women, the infection often seems mild, with flu-like symptoms. However, it can lead to miscarriages, infection of the newborn, or stillbirths.

Listeria is treatable with antibiotics.

Those at risk for *Listeria* should avoid soft cheeses, cook leftovers or ready-to-eat foods until they are very hot, fully cook chicken, and avoid uncooked fish, like sushi.

The safe handling of food and proper hand-washing and surface cleaning, as reviewed earlier in this chapter, is also very important.

> **CAUTION**
>
> **Antigen Alert**
>
> People at highest risk for *Listeria* infection include the following:
>
> ◆ Pregnant women
> ◆ Transplant patients
> ◆ Lymphoma patients
> ◆ AIDS patients
> ◆ Anyone taking cortisone
> ◆ The elderly
> ◆ Dialysis patients
> ◆ Alcoholics

Trichinosis: It's Not Kosher

Eating raw or undercooked meats, especially pork, but also bear, fox, dog, wolf, horse, seal, or walrus, can put you at risk for trichinosis. Trichinosis is caused by eating meat infected with the larvae of a species of worm called *Trichinella*. Infection is most common in areas where raw or undercooked pork, such as ham or sausage, is eaten.

The first symptoms of trichinosis include nausea, diarrhea, vomiting, fatigue, fever, and abdominal discomfort, followed by headaches, fevers, chills, cough, eye swelling, aching joints and muscle pains, itchy skin, diarrhea, or constipation. In severe cases, death can occur.

When a person or animal eats meat that contains *Trichinella* cysts, the acid in the stomach dissolves the hard covering of the cyst and releases the worms. The worms pass into the small intestine and mature in a couple of days. After mating, adult females lay eggs, which develop into immature worms. They travel through the circulatory system to the muscles where they become enclosed in a capsule again.

Prevention requires that meat products be cooked until the juices run clear or to an internal temperature of 180° F. Pork should be frozen for 20 days to kill any worms. The worms in game meats may not effectively be killed by freezing.

Several safe and effective prescription drugs are available to treat trichinosis.

Curing (salting), drying, smoking, or microwaving meat does not consistently kill infectious worms.

Preventing Food-Borne Illness

Proper food safety, proper hand-washing, and proper cooking can prevent most cases of food-borne illness. It's important to do this at home, but it's also important to be careful in restaurants. There is nothing wrong with sending back food that seems to be undercooked or improperly prepared. Doing that may save you from an unpleasant and sometimes serious infection.

The Least You Need to Know

- Many different bacterial organisms can cause food-borne illnesses.
- Proper handling of food, proper hand-washing, and proper cleaning of the kitchen can prevent many infections.
- Early diagnosis is important for proper treatment in most food-borne illnesses.
- Although some long-term effects can occur, most food-borne illness is curable.

Chapter 18

Rare and Terrifying: Ebola and Other Dangerous Diseases

In This Chapter

◆ Sifting through the media hype

◆ Why Ebola is so terrifying

◆ The relationship between monkeys and Marburg

◆ Hanta's presence in the United States

◆ CJD, the human form of mad cow

◆ Tracking Legionnaires' disease

◆ Protecting yourself from rabies

The media loves a good story—one that's full of gory or otherwise disturbing details and that preys upon people's fears. Outbreaks of diseases such as Ebola, Marburg, and mad cow fit this bill perfectly.

Whereas tuberculosis and hepatitis, diseases that cause far more deaths and sicknesses, rarely get a headline, a single outbreak of any of the diseases discussed in this chapter will almost always make the front page of your local daily and get a mention (with lots of frightening footage) on the nightly news.

Blood Gushing Out of Every Orifice: Hemorrhagic Fevers

His eyes are the color of rubies, and his face is an expressionless mass of bruises. The red spots, which a few days before had started out as starlike speckles, have expanded and merged into huge, spontaneous purple shadows: His whole head is turning black-and-blue. The muscles of his face droop. The connective tissue in his face is dissolving, and his face appears to hang from the underlying bone, as if the face is detaching itself from the skull. He opens his mouth and gasps into the bag, and the vomiting goes on endlessly. It will not stop, and he keeps bringing up liquid, long after his stomach should have been empty ….

So goes Richard Preston's description of an Ebola sufferer in his account of the Ebola outbreaks in Zaire, *The Hot Zone*. Ebola is a *hemorrhagic* disease or fever, meaning it is characterized by bleeding. Other hemorrhagic diseases include Marburg and hantavirus (discussed later in this chapter).

These diseases are very contagious and are passed through contact with the blood or body fluids of someone who is infected, or by touching a contaminated surface. Fortunately, neither Ebola nor Marburg have been seen in the United States yet.

Disease Diction

A **hemorrhagic fever** is one characterized by severe bleeding, often from the mouth, stomach, and digestive tract. The term comes from the Greek *Haima* meaning blood and *rhēg-nynai* meaning to burst forth.

Antigen Alert

A closely related virus, called Reston, was isolated from animals in the Philippines, indicating that these diseases are not completely confined to Africa. So far, Reston has not been found to cause disease in people.

Ebola: Africa's Bloody Disease

The Ebola virus was first associated with an outbreak of 318 cases of a hemorrhagic disease in Zaire. Of the 318 cases, 280 of them died—and died quickly. That same year, 1976, 284 people in Sudan also became infected with the virus and 156 died.

The viruses that cause Ebola and Marburg are similar, infecting both monkeys and people. The outbreaks of these diseases are often self-contained, however, because they kill their hosts so quickly that they rapidly run out of people to infect.

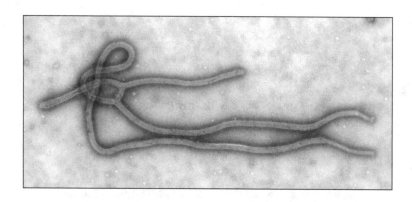

Transmission electron micrograph of the Ebola virus.

(Courtesy CDC. Photo C. Goldsmith)

The Zaire strain of Ebola virus has a mortality rate of 88 percent, which is higher than either the Sudan strain of Ebola or the Marburg virus.

The Ebola virus spreads through the blood, multiplying in many organs. It causes severe damage to the liver, lymphatic system, kidneys, ovaries, and testes. Platelets and linings of arteries are severely damaged, which results in profuse bleeding. Mucosal surfaces of the stomach, heart membrane, and vagina are also affected. Internal bleeding results in shock and acute respiratory distress, leading to death.

Ebola's Frightening Symptoms

Once a patient is infected with Ebola, the incubation period is 4 to 16 days. The onset of disease is sudden, with fever, chills, headache, anorexia, and muscle pain. As the disease progresses, nausea, vomiting, sore throat, stomach pain, and diarrhea are common. Most patients develop severe hemorrhages, usually between days five and seven. Bleeding occurs from multiple sites, including the digestive tract, lungs, and gums. Death occurs within 7 to 16 days.

How Ebola Is Spread

Epidemics result from person-to-person contact within communities, families, and hospitals, or from inadvertent laboratory exposures. The means of infection and the natural ecology of these viruses are largely unknown, although an association with monkeys and/or bats has been suggested.

> **Infectious Knowledge**
>
> The Ebola virus was named after the Ebola River, in the Congo. The disease has occurred in Congo, Sudan, the Ivory Coast, and Uganda. There has never been a case in the United States.

Potent Fact

Some people recover from Ebola. No one fully understands why.

Diagnosing Hemorrhagic Fevers

Marburg, Ebola, and Reston viruses can be isolated from the blood of people who are infected. They produce distinct antibody responses that are easy to detect. These antibodies appear 10 to 14 days after initial infection. Unlike most other infectious agents, there is significant risk to health-care workers and laboratory technicians in handling and processing blood or tissue samples infected with Ebola.

Treatment and Vaccines

There are no drugs that work against these diseases, and there is no known vaccine. Therapy involving blood plasma from people who have recovered, anticoagulation agents, which attempt to reduce hemorrhaging, and interferon have been used with limited success. Their effectiveness remains controversial.

Disease Diction

The **index case** is the first case of disease in an outbreak. If epidemiologists can find the index case, sometimes it helps them figure out how a disease began to spread.

The most effective way to reduce or prevent transmission in an outbreak is through the proper use of barrier protection for doctors and nurses. This includes the use of gloves and masks, with gloves being changed after every patient. Another important protection in a hospital is being sure all equipment is properly sterilized.

People don't carry the virus. They get sick and their infected blood and bodily fluids infect others.

Marburg: Not Just a German City

The Marburg virus was first identified in laboratory workers in Marburg, Germany, and Belgrade, Yugoslavia, in 1967. The researchers contracted the disease from African green monkeys that had been imported from Uganda. Twenty-five cases resulted from direct contact with monkeys (primary cases) and six cases resulted from others' contact with primary cases (secondary cases) in the first outbreak. Seven people died. Since then, there have been sporadic outbreaks in Zimbabwe, South Africa, and Kenya.

The symptoms and course of the disease are very similar to Ebola, although the fatality rate is lower, at approximately 25 percent.

Researchers who study these diseases do so under very strict safety conditions. In fact, there are only a couple of laboratories in the United States that have facilities that are set up to do work on such dangerous and highly contagious organisms.

Hantavirus: Four Corners, United States

In 1993 there was an outbreak of disease in the Four Corners of the United States—where New Mexico, Arizona, Colorado, and Utah all share a border. The illness was unexplained, it occurred primarily in Native Americans, and it was characterized by acute respiratory symptoms—throat irritation, coughing, and sneezing. Thirteen people died in the 1993 outbreak.

It turned out that heavy rains in the spring had increased the food supply for rodents, and that they were spreading the disease to people. The disease was later identified as hantavirus.

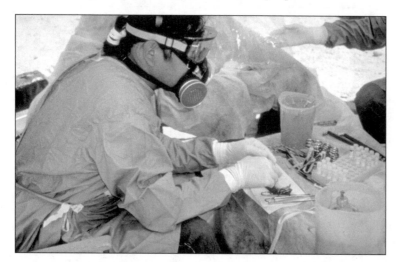

CDC scientist conducting hantavirus field studies by collecting specimens from trapped rodents (1993).

(Courtesy CDC/Cheryl Tryon)

Hantavirus is an RNA virus that is carried by mice and rats. Initial symptoms are nonspecific, like fever, chills, and muscle aches. The cough and severe symptoms of the disease don't usually happen until seven days after initial symptoms. Once they start, the disease progresses quickly with hospitalization and ventilation often required within 24 hours.

Set Your Mouse Traps!

The deer mouse and the cotton rat are the primary carriers of hantavirus. They shed the virus in their urine, droppings, and saliva. When these droppings are stirred up, tiny droplets of virus get into the air; this process is called *aerosolization*. People get the disease when they breathe air that has been contaminated by the virus. The disease is not spread from person to person, nor can it be transmitted from a blood transfusion.

Disease Diction

Aerosolization is when tiny particles of a disease-causing organism become airborne and can cause disease in those who breathe them in. Hantavirus and tuberculosis are spread this way.

A deer mouse, the major carrier of a hantavirus that causes hantavirus pulmonary syndrome in humans.

(Courtesy CDC/NCID)

Hantavirus can also be spread if someone is bitten by a rodent, which is rare, or by touching a surface that is contaminated with the virus and then touching their nose or mouth.

Disinfecting rodent-infested areas is one of the keys to preventing hantavirus. These areas include barns, sheds, warehouses, summer homes closed for the winter, or even homes people live in all year round in areas with large rodent populations.

Who Is at Risk?

People who work around barns, garages, and warehouses are at greatest risk for contracting hantavirus because of their potential exposure or proximity to mouse droppings. Hikers and campers may be at risk as well, especially if they explore vacant buildings. Depending on where you live, you can even be at some risk in your own home.

Hantavirus is not common, but it has been found in about half of the states in this country.

Signs of Hanta

Symptoms usually appear one to six weeks after initial exposure to the virus. The disease, hantavirus pulmonary syndrome, starts out with fatigue, fever, and muscle aches. About half of the people who get sick also have headaches, dizziness, chills, nausea, vomiting, diarrhea, and stomach pain. Seven to 10 days after the first symptoms appear, coughing and shortness of breath begin. This is because the lungs are filled with fluid. At this point the disease becomes much more severe and often requires hospitalization.

Right now there is no treatment or cure for hantavirus pulmonary syndrome. People with advanced cases of the disease should be kept in intensive care and given oxygen to help them breathe.

Hanta's History

Hantavirus was first identified during the Korean War, in the early 1950s. Three thousand United States and United Nations soldiers became infected. The disease got its name because it was found near the Hantaan River in Korea. By October 1995, there were 119 confirmed cases of the disease in the United States; half of those people died.

The disease also occurs in Europe and Asia, but there it is caused by a similar virus that is not quite as deadly as the one that we have in the United States. Worldwide, 100,000 to 200,000 people are infected with this milder strain each year.

> **Infectious Knowledge**
>
> The mice and rats that carry hantavirus don't get sick from it. They only serve as vectors, carrying it and passing it on to people.

Mad Cow Disease: Crazy Cows

Mad cow disease kills cows and some of the people who eat the meat of infected cows. The disease has spread widely in Europe, and many people fear that it could spread to the United States, although so far it hasn't. The disease is caused by a rogue protein, called a prion. Prions are smaller than viruses and are not killed by the heat and chemicals that kill bacteria and viruses, making them particularly difficult organisms to fight.

The disease fills the cow's brain with holes, making it look like a sponge. An infected cow will show changes in temperament that may include nervousness, aggression, abnormal posture, poor coordination, trouble getting up, decreased milk production, and weight loss. There is no treatment for the disease, and all cows that get it will die. The incubation period is between two and eight years, and it takes two weeks to six months for cows to die after symptoms appear.

Countries that have had recent cases of mad cow disease include the Czech Republic, Ireland, Northern Ireland, France, Germany, Greece, Portugal, Slovakia, Slovenia, Spain, Switzerland, Netherlands, Belgium, Denmark, Luxembourg, Liechtenstein, Italy, and Japan.

There is no lab test to detect the mad cow disease in live animals. It is confirmed only during

Potent Fact

Although only cows can contract mad cow disease, humans can be infected with the human form, called Creutzfeld-Jakob, from eating tainted beef.

Antigen Alert

The U.S. Department of Agriculture conducts strict surveillance of livestock to ensure against the transmission of mad cow disease to the United States.

autopsy by microscopic examination of brain tissue and molecular detection of the prion that causes the diseases.

Sheep, humans, cows, elk, deer, mink, rats, mice, hamsters, and possibly monkeys get a variation of the disease.

Infectious Knowledge

In 1982, researcher Dr. Stanley Pruisner, a biochemist at the University of California at San Francisco, identified prions as being responsible for causing mad cow and the human version of the disease, Creutzfeld-Jakob disease. Although his work is still controversial, it is now fairly widely accepted that prions exist and cause these diseases. In 1997, Dr. Pruisner received a Nobel Prize for his work.

Britain's Battle with Mad Cow

England saw its first cases of mad cow disease in 1986. Since then, more than 170,000 cows have been diagnosed with the deadly disease. The British government has spent $7.5 billion to kill and dispose of 4.7 million cows that were old enough to develop the illness.

More than 80 people died in Britain from eating mad cow–tainted meat. Most were infected even as the government assured the public that the disease couldn't infect them. At the time the initial cases occurred, cow by-products were fed to other cows, contributing to the spread of the disease. This practice was stopped in 1989.

CJD: Mad Cow's Human Counterpart

Creutzfeld-Jakob disease (CJD) is a rare, fatal brain disease that causes rapid, progressive dementia and associated neuromuscular disturbances. It is a horrible disease that leads to a gruesome death. People experience mood swings, numbness, uncontrollable body movements, and other neurological symptoms. It is similar to Alzheimers, another disease that affects the brain.

Disease Diction

Creutzfeld-Jakob dis-ease is a rare disease—only one person in a million gets it each year.

The course of the disease takes around 18 months from the first appearance of symptoms to death, although it can be faster in some cases. There is no treatment, and the disease is always fatal. There are three ways people can get CJD: 1) it can occur sporadically; 2) it can be inherited; and 3) it can be transmitted through infection.

Even when it occurs by infection, CJD isn't contagious in the usual sense. Only a few cases have arisen after exposure to an infected individual. Infection of this type occurs as an accidental consequence of a medical procedure that uses tainted human matter or surgical instruments. This can happen with corneal transplants, implantation of electrodes in the brain, or contaminated surgical instruments. Normal human contact is not enough to spread the disease.

From Subtle to Severe

There are several common symptoms in CJD patients. The initial stages can be subtle, with symptoms like insomnia, depression, confusion, personality changes, strange physical sensations, and problems with memory, coordination, and sight. As the disease progresses, people experience rapid, progressive dementia and involuntary jerking movements. Problems with language, sight, muscle weakness, and coordination worsen. In the final stages of the disease, the person loses all mental and physical function, lapses into a coma, and usually dies from an infection.

When to Suspect CJD

CJD should be considered when an adult patient develops rapid dementia and involuntary muscle jerks. Traditional lab tests have not worked to diagnose the disease, unfortunately. MRIs and x-rays have not been helpful, and cerebrospinal fluid usually appears normal. The most helpful test is the electroencephalogram (EEG), which measures brain wave activity and often shows an abnormal pattern when CJD is present.

A definitive diagnosis requires a brain biopsy or autopsy that can detect the changes in brain tissue caused by the disease. Work is ongoing to develop new tests, including ways to identify the prion protein thought to cause the disease.

Like mad cow, there is no effective treatment or cure for CJD. It is 100 percent fatal.

A New Variation of CJD

Creutzfeld-Jakob was first diagnosed in the 1920s. During the past 10 years, a variation of the disease has been found in Great Britain. The average age of patients was 28 as opposed to 63, and the disease duration was 14 months as opposed to four to six months. There have been 10 confirmed cases, and the most likely explanation is a direct link to exposure to infected meat. There have been four cases of this variant of the disease in France, one in Ireland, and none to date in the United States.

Rabies

Rabies is a disease caused by a virus that is fund in the saliva of infected animals. It can affect all mammals. Most of the time, it is transmitted to pets and people by bites, but open cuts can become contaminated as well. If rabies is not identified and treated, it causes painful death. Today, most rabies cases in the United States occur in wild animals like raccoons and bats. Pets account for less than 10 percent of reported cases.

The rabies virus attacks the central nervous system. Unfortunately, early symptoms of rabies in people are general things like fever and headache. As the disease progresses, neurological symptoms show up. They include insomnia, anxiety, confusion, paralysis, excitation, hallucinations, agitation, hypersalivation, difficulty swallowing, and fear of water. People who contract rabies die within days of the appearance of these symptoms.

> **Infectious Knowledge**
>
> According to the CDC, more than 90 percent of rabies cases in animals occur in wildlife. Before 1960, most were in domestic animals.

Very few people in the United States die of rabies. This is due in part to public health measures that include vaccination of pets, animal control programs, good laboratories, and a rabies vaccine that can be given to those at risk as well as to those who have been bitten.

Animal Bites Can Be Deadly!

Rabies is transmitted when an infected animal bites another animal or person, passing the virus in saliva. After initial infection, the virus enters a phase where it is difficult to detect. This phase can last for several days or several months. In some cases, the newly infected animal or person develops an effective immune response and doesn't get sick. In other cases, the virus is transported to the central nervous system.

The incubation period for rabies can be several days or several years, but it is normally one to three months. Once the virus gets into the central nervous system, it acts quickly to cause disease. This is when the previously described symptoms appear. After that, the infected animal or person will die within days.

Testing for Rabies

In animals, the best diagnostic test uses brain tissue, and it can only be done after the animal dies. In people, there are several tests that are used. They require saliva, blood, spinal fluid, and skin biopsies. Antibody tests as well as DNA tests are done.

> **Potent Fact**
>
> In 2000, there were 7,369 cases of rabies in the United States. None of these were in people. 93 percent of the cases were in wild animals.

A diagnostic laboratory takes only a few hours to figure out whether an animal is rabid. This is important when a person has been bitten and is worried about exposure.

By the time symptoms show up, there is no treatment for rabies. However, there is a good rabies vaccine that can be used for people who may be at risk. The same vaccine is also effective after exposure. Domestic pets are also given rabies vaccines.

What If I Think I've Been Exposed?

If you think you have been bitten by an animal that might be rabid, the first thing to do is wash the wound carefully, then call your doctor right away! Your doctor will figure out whether you need to have a rabies vaccine.

If you are a pet owner, be sure to keep vaccinations up to date for your dogs, cats, and ferrets (if you happen to have ferrets). This will protect your pets and it may even help to protect you! If your pet is bitten by a wild animal, get them to the veterinarian immediately. If there are stray or wild animals in your neighborhood, call animal control so that they are removed. Don't let your kids play with them or pet them!

Antigen Alert

What if you're traveling abroad? Rabies is common in certain countries in Asia, Africa, and Latin America. It causes tens of thousands of deaths a year. Before you travel, talk to your doctor about risks and see whether you should receive a rabies vaccine.

Legionnaires' Disease: A Deadly Convention

In the bicentennial summer of 1976, the American Legion symbolically held its fifty-eighth annual convention in Philadelphia at the landmark Bellevue-Stratford Hotel. The eagerly anticipated event began with its usual fanfare and high spirits. But the convention mood turned dour as one veteran after another became deathly ill.

Within a short period, 221 people were stricken with an unknown respiratory syndrome characterized by high fevers, rigors, extreme exhaustion, and poor response to antibiotic therapy. At its conclusion, the epidemic resulted in 34 deaths due to rapidly progressive pneumonia and related complications. The cause of the epidemic baffled the medical and public health community since it did not appear bacterial in origin. Some suggested it was the beginning of an influenza pandemic, since "Swine Flu" had already started infecting large populations in Asia that year. Other speculation ranged from chemical agents to covert conspiracies directed against the American veterans.

Federal and State epidemiologists were investigating the hotel as the epidemic's source because the outbreak was confined, while scientists looked for clues in diseased tissue and bodily fluids. For

Infectious Knowledge

Business at the majestic Bellevue-Stratford Hotel was hurt badly by the devastating summer of 1976. After struggling to operate in the aftermath of the epidemic, it closed in 1979. Over the next decade it opened, closed, and changed names and ownership several times. A once proud hotel dating back to 1904 was forced to scramble and redefine itself to survive. Sometimes diseases fell more than people.

months, questions of origin swirled in the media. The puzzle was ultimately solved by CDC scientist Joseph McDade. Surprisingly, it was a bacterium, *Legionella pneumophila*, aptly named in tribute to the afflicted veterans and as a reflection of the nature of the disease. Once identified, the organism was traced to an air conditioning/cooling system that served the convention hall. Not long after the identification of *Legionella*, it was determined that a similar unexplained episode some eight years earlier in Pontiac, Michigan was caused by the same organism. Cases have now been traced back to 1947.

The Cause

Legionella pneumophila is one of many species of *Legionella*, but only half cause disease in people. The organisms are intracellular pathogens, so cell-mediated immunity plays the most important role in host defense.

Legionella continues to cause minor epidemics throughout the world, and legionellosis is a worldwide problem. Cases appear in sporadic, endemic, and epidemic fashions. Epidemics usually result from exposure to aerosolized bacteria in the hospital or in the workplace. It accounts for one to three percent of community-acquired pneumonias and 13 percent of pneumonias acquired in the hospital.

Other Epidemics and Outbreaks

The second largest outbreak was at the Stafford hospital in England in 1985, where a total of 101 people contracted the disease and 28 died. In late February 1999, an epidemic emerged at a large flower show, the Westfriese Flora (WF) in Bovenkarspel, The Netherlands. Some 181 people developed *Legionella* pneumonia and 21 died. In this case, a whirlpool in one of the main halls was found to be the likely source of infection.

Symptoms of Legionnaires' Disease

Symptoms usually begin 2 to 10 days after a person is infected with the bacterium, but in most cases, symptoms begin after five to six days. Patients develop flu-like symptoms with muscle aches, headache, loss of appetite, and dry cough with persistent high fever. A chest X-ray usually shows pneumonia. A related illness caused by *Legionella*, called Pontiac Fever, consists of fever, headache, weakness, and muscle ache. It is generally less severe, lasts for two to five days, and there is no pneumonia.

How You Can Get It

Legionnaires' disease is acquired by inhaling the *Legionella* bacteria, usually carried as an aerosol mist. It is not spread person to person. It can affect all adults but is most common among the elderly and those with impaired immune systems. Men appear more susceptible than women, and diabetics, smokers, or heavy consumers of alcohol are also at increased risk of getting Legionnaires' disease.

Clinical symptoms are confirmed by growing the bacterium from lung secretions, testing the blood for an immune reaction, or analyzing urine for the presence of the organism. The urine test is most effective three days after symptoms appear.

Potent Fact

Most people exposed to *Legionella* bacteria do not become infected.

Treatment and Prevention

Erythromycin is the antibiotic of choice, although doxycycline has also been used effectively. Exposure can be minimized by inspections of large air conditioning systems with cooling towers and evaporative condensers where the organisms proliferate. Window air conditioners are not a risk for disease transmission.

Where *Legionella* Likes to Live

Water reservoirs are the natural habitat of *Legionella pneumophila*, although the bacteria are found in many environments including soil and dust. The bacteria are also found in hot and cold water taps, showers, whirlpool baths, creeks, ponds, and wet soil. Warm water helps *Legionella* multiply and cases increase in warm weather. Initially, air conditioning cooling towers were implicated as the primary source of infections through aerosols. However, it is generally recognized that contaminated supplies of drinking water are also a common source of spread to patients.

Trojan Horse Theory

It was recently found that amoebae also proliferate in warm waters and provide *Legionella* shelter from unfavorable environmental factors. The exact role of amoebic hosts in bacterial survival, multiplication, and transmission is just emerging. But it has been suggested that once amoeba harboring *Legionella* are ingested, the bacterium quietly emerges from its shelter, where it can infect and cause disease. Thus, the free-living amoebae are the "Trojan horses" of the microbial world.

The Least You Need to Know

◆ There have been no documented cases of Ebola or Marburg in the United States.

◆ Hemorrhagic fevers are characterized by bleeding, and the blood of victims of hemorrhagic fevers is highly contagious.

◆ Sickness from the hantavirus is rare, but the virus has been identified in half of the U.S. states.

◆ There's no cure for CJD, the human form of mad cow disease, and people infected with it always die from it.

◆ Although Legionnaires' disease is rare, outbreaks do still occur.

Chapter **19**

Common Diseases with an Infectious Twist: Ulcers and Viral Cancers

In This Chapter

◆ Diseases that you wouldn't think are infectious

◆ What *really* causes ulcers

◆ Cancers caused by viruses

Bet you always thought you couldn't "catch" cancer. And you never hear about ulcer epidemics, do you? Many people are surprised to learn that certain forms of cancer and ulcers are caused by infectious organisms and can be transmitted just like other infectious diseases. It turns out that 80 percent of all ulcers are actually caused by a bacterium and between 15 and 20 percent of all cancers are caused by viruses.

In this chapter, you will learn about diseases that we don't usually think of as infectious.

Taking the Stress out of Stomach Ulcers

Until recently, ulcers were believed to be caused by excess stomach or intestinal acid, induced by stress or nervousness, spicy food, disrespectful children, the reappearance of long-lost relatives, or poor choices in friends or romantic partners. Within the last 20 years, however, researchers discovered that a microorganism, not emotionally charged situations, causes our peptic distress. It is estimated that 25 million people in the United States have had peptic ulcer disease, and amazingly, a huge percentage of these cases are due to a microorganism called *Helicobacter pylori*.

Infectious Knowledge

In one of the great stories of modern medicine, two physicians from Western Australia, J. Robin Warren and Barry Marshall, isolated the *Helicobacter pylori* bacterium and showed that it was the cause of gastritis and stomach ulcers. Warren, a pathologist, had observed unusual curved and rod-shaped bacteria under the microscope during biopsies of ulcer tissue. He further observed that the more organisms present, the higher the degree of inflammation, and that they occurred in many routine gastric biopsy specimens.

Barry Marshall, an internal medicine student, became interested in this finding and, along with Warren, set out to understand the role of the bacterium in ulcer disease. In 1982, Warren and Marshall cultured these organisms from 11 patients with gastritis. Once they found a way to isolate and grow the organism, Marshall and an assistant embarked on a bold attempt to fulfill Koch's third and fourth postulates by eating cultures of the bacteria to see if they caused disease. In fact, both contracted gastritis (inflammation of stomach lining), and the organism could be cultured from their biopsied tissue. Incredibly, Marshall fulfilled Koch's postulates for the role of *Helicobacter pylori* in gastritis by self-experimentation! Fortunately, he also showed that gastritis could be cured with antibiotics and bismuth salts.

Infections with *Helicobacter pylori*, the microorganism responsible for most stomach ulcers, are very common throughout the world, occurring in 40 to 50 percent of the population in developed countries and in 80 to 90 percent of the population in developing regions. *Helicobacter pylori* infection is low in children, and the infection rate rises dramatically in people over 50 years of age.

Potent Fact

More than half of the world's population over 50 years is infected with *Helicobacter pylori*, the bacterium that causes stomach ulcers.

Got Gas?

Diseases caused by an *Helicobacter pylori* infection include gastritis, dyspepsia, and ulcers. Gastritis and dyspepsia cause discomfort, bloating, nausea, and sometimes vomiting. Ulcers cause burning or pain in the upper abdomen, and usually occur about an hour after meals or at night. The symptoms are often relieved temporarily by antacids, milk, or medications that reduce stomach acidity.

Antigen Alert

Besides *Helicobacter pylori*, gastritis can be caused by excessive alcohol consumption and by certain drugs, such as aspirin and ibuprofen.

How *Helicobacter Pylori* Works

Helicobacter pylori is a rod-shaped bacterium that is found in the protective mucous layer of the stomach. The bacteria have long threads protruding from them that attach to the underlying stomach cells. The mucous layer that protects the stomach cells from acid also protects *Helicobacter pylori*. About one half of the strains of the bacteria produce poisons that help induce gastritis and peptic ulcers.

People with severe gastritis are more likely to be infected with a strain of the bacteria that secretes poisons, which cause local inflammation and damage to the stomach lining, leading to ulcers. Infected individuals who show no symptoms probably have a strain that does not secrete poisons.

Gastric Cancer

Gastric cancer is one of the most deadly forms of cancer, and there is evidence that it is initiated by an infection from the *Helicobacter pylori* bacteria. The most important evidence for this connection is a study from 1993 that involved people from 13 different countries. The study showed that populations with a 100 percent *Helicobacter pylori* infection rate (meaning that everyone had the bacterium) were six times more likely to develop gastric cancer as a similar population with no infection. As a result, *Helicobacter pylori* infection has been classified as a carcinogen—cancer-causing agent—by the World Health Organization.

Disease Diction

An **endoscopy** is a procedure where an endoscope (a flexible tube with an optical system attached) is inserted through the mouth or anus to view the interior of the body. Endoscopes can also be used to take tissue samples that can be cultured to diagnose a disease as well as to perform minor surgical procedures like treatment of bleeding lesions or removal of colon polyps.

Testing for *Helicobacter Pylori*

Our body's immune system engages in a systemic response to *Helicobacter pylori* infection, which means that antibody tests can be used to show its presence. However, cultures, which can be obtained through *endoscopy*, are generally regarded as the best way to detect the bacterium.

A breath test also can be used to detect *Helicobacter pylori*, which is helpful in both diagnosing and confirming the eradication of the bacteria following treatment.

Most people who are infected with the *Helicobacter pylori* bacterium have no symptoms and don't need to be treated. For people with symptoms, the most effective treatment is to reduce stomach acidity with antibiotics or other drugs. This type of treatment results in eradication rates and ulcer healing in greater than 80 to 90 percent of cases, with relapse rates of less than 10 percent. By prescribing a combination of antibiotics, called combination antibiotic therapy, physicians can help minimize the development of drug-resistant strains of the bacteria.

> **Infectious Knowledge**
>
> An oral vaccine for *Helicobacter pylori* is currently in clinical trials.

Viral Cancers

Cancer is a disease involving altered genes and altered gene function. The gene mutations cause cells to grow out of control and form tumors, or masses. If the tumors aren't removed, or even if they are and just one or two cancerous cells remain, the disease comes back and, in some cases, causes death. Although we usually think of cancer as an inherited disease or one caused by chemical or physical damages, like smoking or exposure to certain chemicals, some types are caused by viruses.

As a matter of fact, viruses are responsible for between 15 and 20 percent of all cancers in people. From a public health perspective, the idea that malignant tumors can sometimes be categorized as a transmissible disease is significant and disturbing.

> **CAUTION**
>
> **Antigen Alert**
>
> Cervical cancer is the second most common cancer in women worldwide. In the United States, there are 13,000 cases per year and 5,000 deaths. Infection with human papilloma virus is the most significant risk factor.

How Does a Virus Cause Cancer?

Some viruses can change the genetic makeup of cells and cause them to become cancerous—to grow out of control. The chances of this happening are greater in someone with a weakened immune system, because the immune system will be less likely to stop tumor growth.

Many viral cancers produce antigens, so a T cell immune response that is antigen-specific could potentially kill the tumor cells. There is significant research into using this concept to try to develop vaccines that could cause the body to launch an immune response to kill viral tumors.

Common Viral Cancers

The most common viruses that can cause cancer are the Epstein-Barr virus, the human papilloma virus (HPV), and the hepatitis C virus. The Epstein-Barr virus is most often associated with Burkitt's lymphoma (a rare form of cancer, often of the face and jaw), cancer of the nose and throat, stomach cancers, and certain B cell lymphomas (any form of cancer that involves the malignant growth of B cells) in patients with weakened immune systems. (HPV is discussed in Chapter 10. See Chapter 11 for more about hepatitis C.)

As researchers have learned, it is often difficult to pinpoint the cause of a disease, and many diseases, such as cancer, can have multiple causes, making them difficult to treat and cure. The ulcer example points out that as old diseases reemerge and new diseases are found, it is important to remember to look for infectious causes, even if they don't seem likely or obvious at first.

The Least You Need to Know

- Most stomach and gastrointestinal ulcers are caused by the bacteria *Helicobacter pylori*, rather than stress or emotional upset, and can be effectively treated or eradicated.
- A link has been found between the presence of *Helicobacter pylori* and some forms of gastric cancer.
- Viruses cause between 15 and 20 percent of all human cancers.

Part 4

Scourges of Our Own Creation

Dr. Frankenstein didn't mean to create a monster and, like the unfortunate doctor, humans didn't intend to make most of the diseases discussed in this section. But as you've probably figured out by now, infectious organisms are opportunistic little guys. You give 'em an inch, they take a mile. So it is that we have disease-resistant strains of many infectious diseases. Hospitals are particularly rich environments for drug-resistant organisms, including some staph infections, which are resistant to all but one antibiotic, and enterococcal infections, some of which are resistant to even our strongest drugs.

Not all infectious diseases and epidemics are unhappy accidents, though. People have purposefully used infectious diseases for centuries to defeat their enemies and wipe out entire groups of people. In this century, countries including the United States, Japan, the Soviet Union, and Britain have tried their hand at engineering bioweapons, creating strains of infectious organisms that are particularly deadly and easily transmitted. And as Americans are all too aware, the threat of innocent people being exposed to such deadly man-made organisms is frighteningly real.

Chapter 20

Bioterrorism: Weaponizing Nature

In This Chapter

- ◆ The history of bioterrorism
- ◆ Potential agents of terror
- ◆ Anthrax attacks in the United States
- ◆ Preparing for another attack

Almost everyone can agree that the notion of "catching" any of the infectious diseases discussed in this book is frightening. But when you factor in the idea of someone—or some group or country—intentionally spreading infectious diseases such as anthrax or the plague, things get downright terrifying. Unfortunately, Americans are all too familiar with this terror. Although the anthrax attacks on the United States in the fall of 2001 only infected a relatively small number of people, they terrorized millions.

In this chapter you'll learn about how disease-causing biological organisms can be used in warfare. You'll find out what it means to "weaponize" an organism, how such organisms have been used in the past, and what organisms are ideally suited for such nefarious uses.

Bioweapons

Biological warfare and bioterrorism are the intentional use of disease-causing organisms—bacteria, viruses, or fungi—as weapons to cause serious illness, resulting in disabling physical and psychological trauma and/or death. The CDC categorizes biologic agents that have the potential to be used as weapons based on several factors, including their potential for widespread transmission, high death rates, and causing public panic and social disruption.

The organisms of highest concern are those that cause …

- **Anthrax** (discussed later in this chapter).
- **Bubonic plague** (see Chapter 6).
- **Smallpox** (see Chapter 6).
- **Botulism** (discussed later in this chapter).
- **Tularemia** (discussed later in this chapter).
- **Ebola or Marburg hemorrhagic fever** (see Chapter 18).
- **Lassa fever.**
- **Multi-drug-resistant tuberculosis** (see Chapter 8).

> **Infectious Knowledge**
>
> Not all infectious organisms are good candidates for biowarfare. In order to be effective, the disease must be readily transmissible and cause high morbidity and mortality.

Historically, bioterrorism has relied on naturally occurring agents. Regrettably, as technology has advanced, it has become easier to produce drug- and vaccine-resistant forms of disease-causing organisms. The prospect for such "smart biological weapons" is scary and daunting.

> **CAUTION**
>
> **Antigen Alert**
>
> Fears have been expressed for some time that hospital staph infections, which already kill thousands of people each year, may be engineered to be resistant to vancomycin, the only remaining antibiotic available to treat these infections. (See Chapter 21 for more on staph.)

Biological Warfare: A Brief History

The use of disease-causing agents as weapons of war has ancient roots. During the Peloponnesian War, a devastating epidemic broke out that killed thousands of Athenians and helped Sparta win the war. The historian Thucydides reported that Sparta infected the wells of Athens. Whether true or not, Sparta's victory was forever tainted, as it became the first recorded rogue state to employ a biological weapon.

Hannibal, the leader of the army of Carthage, used biological warfare of a different kind. In 190 B.C.E., he won a famous naval battle by hurling jars filled with venomous snakes at the enemy ships. Upon breaking, terrified sailors were thrown into chaos, which cost them their lives.

A Different Kind of Catapult

In 1346, the Tartar army was besieging the city of Kaffa, located in present-day Ukraine. During the onslaught, the Tartars suffered an outbreak of the bubonic plague, which decimated their forces. Not ready to give up, the Tartars tried to infect Kaffa with the same plague that was killing them by catapulting bodies of plague victims over the walls of the town. In fact, the city defenders subsequently contracted bubonic plague and abandoned the city to the Tartars. Although the Tartars took credit for spreading the disease, it was much more likely that fleas from rats in the Tartar camp were responsible for the mass epidemic in Kaffa.

A Pox on the Indians

Perhaps the most sordid use of biological warfare occurred during the French and Indian Wars. Sir Jeffrey Amherst, commander of the British forces, formulated a plan to reduce the size of the Native American tribes in the Ohio Valley by infecting them with smallpox.

In late spring 1763, his men collected blankets and a handkerchief laden with the pus or dried scabs from the smallpox sores of the British troops suffering from smallpox. They gave the blankets and handkerchiefs to the Indians as a gesture of good will. As expected, the Native American tribes, who had never been exposed to smallpox and so had developed no immunity to it, suffered a devastating smallpox epidemic.

The Modern Era of Biological Warfare

Unfortunately, biological warfare isn't restricted to the distant past. Modern examples use more sophisticated techniques to spread organisms, and have the potential to cause damage over large geographic areas.

In other words, modern biological warfare has taken a nasty turn: It involves weaponizing infectious organisms. *Weaponizing* means processing a biological agent so it can easily be delivered to harm or kill people. This can be genetic manipulation, such as making drug-resistant strains to spread, or it can be manual, such as drying anthrax so it is more likely to get into the air and cause disease.

Disease Diction

Weaponizing means to take a biological agent and process it so that it can be easily delivered to harm or kill people.

Weaponized or not, organisms can be spread around many ways. For example, they can be sprayed, placed in bombs, sent through the mail, or put in ventilation systems.

Japan's Unit 731

Japan's World War II biological warfare program, referred to as Unit 731, was horrific. Unit 731 employed a staff of more than 3,000 scientists and technicians in 150 buildings. Scientists experimented on prisoners by infecting them with dysentery, cholera, and bubonic plague. At least 10,000 Chinese, Korean, and Russian prisoners died.

Potent Fact

The Chinese weren't the only ones who suffered from Japan's biowarfare campaign. Aerial spraying killed 1,700 Japanese troops, too.

The Japanese didn't confine the terror to Unit 731, though. The military sprayed 11 Chinese cities with biological agents, including cholera, anthrax, *Salmonella*, and the plague. Low-flying planes dropped fifteen million fleas from plague-infected rats, and plague-infested rats were released in Chinese cities. Fortunately, Japan's biowarfare campaign wasn't as effective as they hoped; but in 1941, nearly 10,000 cholera cases emerged in the city of Changteh.

Antigen Alert

Gruinard Island is only a few hundred meters from mainland Scotland. Although decontamination hasn't worked, there is still wildlife on the island and people are able to go there if they want. Not surprisingly, neither Frommer's or Fodor's mention Gruinard Island as a daytrip.

The Brits' Biowarfare Program

During this same period (WWII), the British focused on anthrax for their biological warfare program. They tested the effectiveness of using conventional bombs to deliver weaponized anthrax on Gruinard Island, off the coast of Scotland. After repeated bombings, an outbreak of anthrax occurred in sheep and cattle on the coast of Scotland facing Gruinard Island. Attempts to decontaminate the island were ineffective, and it remains contaminated to this day.

The U.S. Takes a Turn

The United States' offensive biological warfare program began in 1942 with major research and development facilities at Camp Detrick, Maryland, where anthrax bombs were produced, and production and testing facilities in Indiana, Mississippi, and Utah.

In a program that was highly classified at the time, the United States military released largely nonpathogenic strains of two disease-causing organisms into large urban areas in order to determine if and how such organisms would spread once released in the environment. In San Francisco, the spraying of *Serratia marcescens*, a microbe that lives in the soil, resulted in at least one death and ten hospital-izations. In 1966, the vulnerability of the New York City subway system was tested by releas-ing a nonpathogenic relative of anthrax, *Bacillus subtilis*, into the subways. The results showed that the release of an organism in one station could infect the entire subway system due to winds and vacuum created by the passing subway trains. The program was ultimately shut down in 1969.

Potent Fact

President Nixon signed National Security Decisions 35 and 44 in November of 1969 and February of 1970 terminating the United States' offensive biologi-cal warfare program.

Infectious Knowledge

The Miracle of Bolsena is depicted on the walls of the Vatican in a painting by Raphael. The German priest Peter of Prague is shown breaking bread for communion at the Church of Saint Christina in Bolsena, Italy. Imagine Peter's surprise when he broke the communion wafer and saw it had blood on it! A true miracle. Or was it?

To honor the miracle, Pope Urban instituted the feast of Corpus Christi (Body of Christ) in 1264. Neither the Pope nor Peter the Priest could ever have known that a red bac-terium, *Serratia marscesens*, a potential agent of bioterrorism, was the probable cause of this blood-like substance on the communion bread. The organism is known to grow on bread that has been stored in a damp place.

Biowar, Soviet-Style

The Soviet Union had its own bioweapons program. Smallpox, anthrax, and tularemia were all mass-produced. They constructed the world's largest biological production plant at a cost of $1 billion. The plant could produce and store 500 tons of anthrax spores, enough bacteria to destroy all urban life on the planet.

In April and May of 1979, an anthrax outbreak occurred in Sverdlovsk, an industrial city of just over a million people, 850 miles east of Moscow. Soviet officials initially attributed the incident to the consumption of anthrax-contaminated meat. However, most of the world's scientific community suspected that the outbreak was caused by the release of anthrax spores from a nearby military facility. Most people who contracted anthrax early in the outbreak lived and worked very near—and downwind from—the military micro-biology lab. Livestock located as far away as 30 miles downwind were also killed.

All told, 94 people developed anthrax disease, and 64 died from inhalation anthrax. Some 13 years later, President Boris Yeltsin admitted that the anthrax outbreak was the result of an unintentional release of spores.

Iraq?

Iraq is suspected of engaging in bioweapons research, including developing weaponized anthrax. So far, the United Nations has not been able to verify such reports, but there is documented evidence that Iraq used chemical weapons during the Iran-Iraq War (1980–1988), resulting in over 2,200 casualties.

America's Anthrax Nightmare

The devastating terrorist attacks on the World Trade Center and the Pentagon on September 11, 2001, left a fragile American public bracing for a new round of unconventional terrorism. Almost immediately, the media began warning of the potential for biological warfare as the next phase of terrorist activity. In early October 2001, this fear was realized when Robert Stevens, a 63-year-old tabloid editor, was diagnosed with inhalation anthrax and died several days later.

Initially, doctors and researchers thought Stevens's infection was an isolated incident, perhaps from environmental exposure—that he had come into contact with naturally-occurring anthrax. However, when his computer keyboard at work tested positive for anthrax spores and nasal swabs of several of his co-workers, including a mailroom worker, tested positive for spores, it became clear that Stevens's exposure was not incidental but deliberate. At about the same time, a powder-filled letter was sent to NBC news anchor Tom Brokaw and was opened by an assistant. Several days later, she was hospitalized with an unusual rash and lesion, which was diagnosed as cutaneous anthrax, a milder form of the disease, which mainly affects the skin.

By mid-October, 12 people were confirmed to have been exposed to anthrax. On October 15, an anthrax-laden letter was opened in the office of United States Senate Majority Leader Tom Daschle. Tens of millions of spores spewed from the letter, heavily contaminating the Hart Senate Office Building. It was now clear that a full-fledged assault was underway, with the U.S. postal system as the messenger of death.

Potent Fact

Anthrax spores can be especially deadly if they enter the alveoli, which are the small air sacks of the lungs. In principle, each anthrax spore is small enough to enter the alveoli. The problem is that spores are sticky and tend to clump, which renders them much less harmful. A major objective of the cold war bioweapons programs in the United States and Russia was to develop procedures for mass-producing spores that would stay dispersed. Chemical treatments with agents such as bentonite and mechanical milling were used to make spores more pathogenic. Weapons-grade anthrax is nearly invisible and highly deadly. More importantly, its presence would be the signature of a military threat.

New cases emerged, especially among U.S. postal employees, who had come into contact with the infected letters. Despite being sealed with tape, spores passed readily through the paper envelopes, resulting in mass contamination of letters and postal facilities. Millions of mail pieces had to be quarantined and the mail facilities sterilized. Health-care and public-health professionals scrambled to learn about the signs, symptoms, and different forms of anthrax infection.

Panicked, hundreds of thousands of people started stockpiling Cipro, the only approved drug for treating anthrax, in case of future attacks. In mid-November, another anthrax letter, with writing identical to that in Senator Daschle's letter, was discovered destined for Sen. Patrick Leahy (Democrat from Vermont).

> **Antigen Alert**
>
> A nasal swab cannot be used to rule out anthrax exposure. Nasal swabs are a form of environmental sampling, and a negative result does not rule out exposure. In fact, nasal swabs of most of the infected people in 2001 came up negative.

For more than two months, new cases emerged. By late November, 11 people had become ill with inhalation anthrax infections and 5 had died. Another 7 people had developed cutaneous anthrax. More than 50,000 people were given antibiotics. By infectious disease standards, the death toll was small. But from the American public's perspective, the threat of bioterrorism was much greater than it had ever been.

A gaseous compound called chlorine dioxide, which effectively kills live organisms and spores, was used to sterilize the Hart Senate Office Building. In this process, the building was sealed and the gas was pumped in for a certain length of time. Once the gas is removed, surface samples were taken to determine the effectiveness of the sterilization treatment. In the case of the Hart building, it took three tries to kill the spores.

Infectious Knowledge

So, who did it? Early speculation pointed to Middle East extremist nations like Iraq. But genetic analysis of the spores quickly indicated that the strain in the letters was the home-grown Ames strain. The FBI leaned toward a domestic terrorist theory and compiled several profiles of individuals suspected of causing the anthrax attack. It was not until a more advanced type of DNA fingerprinting was performed that it was discovered that the anthrax was closely linked to United States military research programs, meaning that the anthrax source was most likely a government lab or a government contractor. The terrorist may be an employee, former employee, or someone with a close link to these facilities. Regrettably, he or she is still at large as of early 2002.

Are We Prepared for Another Attack?

Are we prepared for another round of bioterrorism? Yes and no. A major bioterrorist attack would be met with a better-coordinated effort, provided the pathogenic agent is known and can be dealt with readily.

The CDC has put in place a laboratory and hospital-based response network that can detect and report unusual biological organisms and quickly identify clusters of disease, which indicate a potential outbreak. The nation is also stockpiling antibiotics and vaccines to be mobilized quickly, as needed.

A problem will arise, however, if large numbers of people are quickly exposed in many different parts of the country. Such a well-coordinated attack would overload local hospitals and public health facilities and would dilute the response capability of the federal government to provide support. In addition, if drug and/or vaccine-resistant strains emerge, our ability to respond effectively will be vastly curtailed.

Natural Outbreak vs. Intentional Attack

When distinguishing between a natural outbreak of a disease and an intentional attack, scientists look at the timing and distribution of cases to determine what's called the epidemic curve, which indicates how the disease is spreading. In a naturally occurring outbreak, the number of disease cases increase progressively as a larger number of people come in contact with patients and other sources that spread disease. However, in a bioterrorism attack, there is a single, or point, source with many people encountering the agent at the same time. The number of cases in the epidemic curve would peak in a matter of days in this situation. If the agent is contagious or a second source occurs, such as the second contaminated letter in the 2001 attacks, then a second peak would occur. The epidemic curve for the anthrax outbreak in fall 2001 shows that the outbreak was intentional.

Number of bioterrorism-related anthrax cases, by date of onset and work location—District of Columbia (DC), Florida (FL), New Jersey (NJ), and New York City (NYC), September 16 through October 25, 2001.

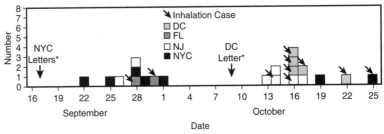

* Postmarked date of known contaminated letters.

All About Anthrax

Bacillus anthracis, the bacterium that causes anthrax, was one of the first organisms shown to cause a disease and was discovered by Robert Koch in 1877.

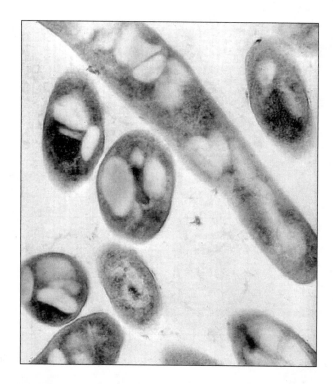

Transmission electron micrograph image of Bacillus anthracis, *the bacterium that causes anthrax.*

(Courtesy CDC/Dr. Sherif Zaki/Elizabeth White)

Anthrax is an organism that typically lives in the soil and commonly infects domesticated and wild animals, such as cattle, sheep, horses, mules, and goats. People become infected when they come into contact with the flesh, bones, hides, hair, or excrement of diseased animals. For this reason, anthrax is fairly common in the textile industry, where it's called wool sorters' disease. Anthrax infections occur sporadically throughout the world. In the United States, areas of infection include South Dakota, Nebraska, Arkansas, Texas, Louisiana, Mississippi, and California.

Anthrax infection results in three different types of disease:

♦ **Cutaneous (skin) anthrax** is initiated when anthrax spores are introduced into the skin through cuts or abrasions. Spores germinate within hours, and the cells multiply and produce anthrax toxin. Initial infection begins as a raised itchy bump that resembles an insect bite but within one to two days develops into a vesicle and then a thick raised escher (sore) with a characteristic black necrotic (dead tissue) center. Lymph glands in the adjacent area may swell. About 20 percent of untreated cases of cutaneous anthrax will result in death.

> **Infectious Knowledge**
>
> The name anthrax refers to the "anthracite-like" blackness of the dead tissue.

Cutaneous anthrax lesions on neck and face of a man.

(Courtesy CDC)

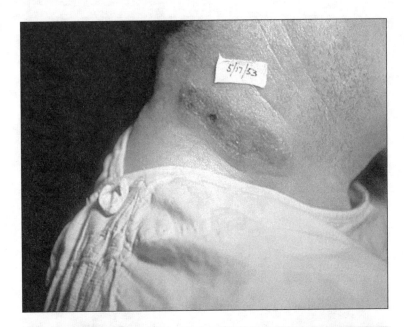

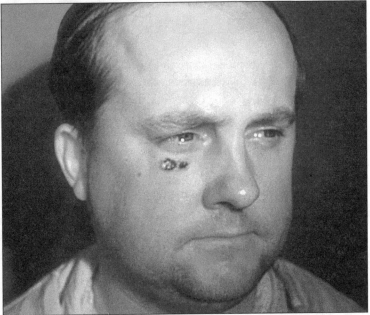

◆ **Gastrointestinal anthrax** occurs after eating contaminated meat. The bacteria move to regional lymph nodes, where they multiply and disseminate. Initial signs of disease include nausea, loss of appetite, and fever followed by acute abdominal pain, vomiting of blood, and severe diarrhea. If untreated, intestinal anthrax results in death in 25 to 60 percent of cases.

♦ **Inhalation anthrax** is the most feared form of the disease. It occurs when small individual spores reach the lungs, where they are transported to lymph nodes. Following germination, when spores begin to grow after a period of dormancy, a large amount of anthrax toxin, or poison, is produced. Patients often have profuse sweating. Parts of the lymph system are overwhelmed and the poison circulates throughout the body. This causes septic shock, a serious condition when blood flow through the body is insufficient resulting in low blood pressure and decreased urine output. Septic shock is a medical emergency and it results in death more than 50 percent of the time.

Potent Fact

Most textbooks say that an infectious dose of anthrax requires 8,000–12,000 spores. However, during the fall 2001 anthrax epidemic, two fatal cases of inhalation anthrax and a case of cutaneous anthrax were recorded in which there was no direct evidence of any spore exposure. The infections were most likely from incidental exposure through mail that had spores on it because it came into contact with other contaminated mail or through an environmental source like soil. The textbooks failed to take into account a subset of the population that was more susceptible to infection because of a weakened immune system.

The presence of anthrax can be determined by growing cultures of live organisms from specimens such as blood, tissue, and sputum. In addition, molecular-based techniques can be used to rapidly identify the organism.

Treating Anthrax

Once anthrax is diagnosed or suspected, especially the inhalation form, it should be treated aggressively. Antibiotics such as penicillin, tetracyclines, and fluoroquinolones are effective if administered within the first 24 hours of exposure. Fortunately, there is little evidence that anthrax has developed resistance to penicillin.

Inhalation anthrax was formerly thought to be nearly 100 percent fatal, despite antibiotic treatment, particularly if treatment was started after symptoms appear. But six cases of inhalation anthrax in fall 2001 were successfully treated by aggressive therapy with the drug Cipro.

The Anthrax Vaccine

An effective vaccine for anthrax exposure is available. Even though 30 percent of the people who receive the vaccine have minor reactions at the injection site, serious side effects are rare. Six inoculations are required for the vaccine to be effective: three within the first two weeks and three more over the next eighteen months. An annual booster shot is required to maintain immunity. The vaccine should only be administered to healthy men and women; pregnant women should not receive it.

Tularemia: Rabbit Fever

The bacterium that causes tularemia, or rabbit fever, is one of the most infectious disease-causing agents known, requiring only 10 organisms to cause disease. It is so infectious that just examining an open culture plate in a laboratory can cause infection. It is a suitable biological weapon because of its strength, ease of dissemination, and potential to cause widespread illness and death.

Tularemia was part of Japan's World War II bioweapons program, and was stockpiled by both the former Soviet Union and the U.S. military. A World Health Organization (WHO) committee reported in 1970 that if 110 pounds of virulent bacteria was spread as an aerosol over a metropolitan area with a population of 5 million, there would be approximately 250,000 illnesses and 19,000 deaths.

Type A or Type B?

The bacterium that causes tularemia is capable of surviving for a long time in water, soil, hay, or decaying animal carcasses. Two main types are known:

- **Type A** is most common in North America and is considered highly virulent in humans and animals.
- **Type B** causes human tularemia, and is common in Europe and Asia. Type B is not as virulent as Type A and is unlikely to cause death.

The overall mortality rate for severe Type A strains is 5 to 15 percent, but may be as high as 30 to 60 percent without treatment. The mortality rate is typically about 2 percent with treatment.

The symptoms of tularemia vary dramatically, depending on the site of infection. In most cases, an infected person will have fever, chills, and head and body aches. Other potential symptoms include skin lesions with swollen glands, swollen glands with no skin lesions, pink eye with swollen glands, sore throat and swollen glands, intestinal pain, diarrhea, vomiting, pneumonia, and fever and systemic illness. Complications include meningitis and severe pneumonia, among others. In a bioterrorism attack with aerosol organisms, fever and pneumonia would be expected.

Rabbits and Rodents

Tularemia is found naturally in small mammals, mice, water rats, squirrels, rabbits, and hares. Sporadic human infections occur through bites, handling infected animal tissues or fluids, inhaling dust from contaminated soil, or handling contaminated pelts or paws of animals. Although many wild and domestic animals can be infected, the rabbit is most often involved in disease outbreaks (hence, its name).

Humans can contract the disease from direct contact with an infected animal, or from the bite of an infected flea or tick. Infection may also occur by ingestion of contaminated water, food, or soil, and inhalation of infective aerosols. There is no documented evidence that the disease can be transmitted from one human to another.

The organism that causes tularemia is found in rural areas. It is much more common in animals than it is in people. The disease is rare in the United States, with only about 200 cases per year. These cases occur mostly in the south-central and western states, in rural areas.

Reading the Signs

Diagnosing tularemia can be difficult because the symptoms vary so much. If tularemia is suspected, tests can be done on the patient's blood, ulcers, eye secretions, sputum, lung fluid, lymph nodes, stomach acid, and throat secretions, or through a blood test for antibodies. Significant numbers of antibodies don't appear in the blood until the end of the second week of illness; the presence of antibodies peaks at four to five weeks, and they can remain present in the system for more than ten years.

Antibiotics such as streptomycin and tetracycline are effective in treating tularemia. Others, such as gentamycin and tobramycin, may also be effective. In a bioterrorism attack, prompt treatment with streptomycin, gentamicin, doxycycline, or cipro-floxacin is recommended. Preventive use of doxy-cycline or ciprofloxacin may also be useful immediately after exposure.

> **Infectious Knowledge**
>
> A vaccine for tularemia has been used to protect laboratory personnel, but given the short incubation period of tularemia (it takes only three to five days from infection for symptoms to appear) and incomplete protection of current vaccines against tularemia, vaccination is not recommended for the general population.

Botulism: Not Just Food Poisoning

During 1895 in Ellezelles, Belgium, three people died from food poisoning, and within a few days, others became seriously ill. A thorough investigation by E. van Ermengem led to the discovery of *Clostridium botulinum*, the bacterium that causes the disease, as well as the discovery of the botulinum toxin, the poison the bacteria secretes that makes us sick. (See Chapter 17 for information on other food-borne organisms.)

Approximately six percent of people who are infected with botulism die. Since 1973, about 24 cases of food-borne botulism, three cases of botulism from wounds, and 71 cases of botulism in infants have been reported annually to the Centers for Disease Control and Prevention (CDC).

The potency of the toxin produced by botulism (called the botulinum toxin) makes it an ideal weapon of biological warfare; terrorists have already attempted to use botulinum toxin as a bioweapon. It is one of the most potent biological toxins known. The Japanese cult Aum Shinriki sprayed the toxin at multiple sites in Tokyo, Japan, and at United States military installations in Japan on at least three separate occasions between 1990 and 1995. Fortunately, they could not deliver the toxin in a way that caused disease. Iraq admitted after the 1991 Persian Gulf War to having produced 19,000 liters (more than 4,500 gallons) of concentrated botulinum toxin. Approximately 10,000 liters (about 2,500 gallons) were loaded into military weapons.

A Paralyzing Meal

Botulism is characterized by paralysis of nerves that happens on both sides of the body. It starts in the head and moves downward. Blurred vision and impaired speech are common initial symptoms. Symptoms of food-borne botulism may begin as early as two hours or as long as eight days after ingestion of toxin. Typically, symptoms are present 12 to 72 hours after the infected meal.

Killing the Poison

Botulinum toxin is colorless, odorless, and tasteless. Fortunately, the toxin can be easily killed by heating, which means that the food-borne botulism is always transmitted by foods that are not heated properly before eating. All forms of botulism result from absorption of botulinum toxin into the circulation from either a mucosal surface such as the gut and lungs, or a wound. Botulinum toxin does not penetrate intact skin. The toxin can be turned into an aerosol spray (which is what the Aum Shinriki cult did) to produce inhalational botulism.

Diagnosing botulism starts with testing suspected contaminated foods and ruling out other causes of neurologic dysfunction that mimic botulism, such as stroke and Guillain-Barré Syndrome (see Chapter 17 for a description of Guillain-Barré). Only a few state laboratories and the CDC perform the tests needed to detect botulinum neurotoxin and bacterial cells in the patient samples and in suspected food samples.

Infectious Knowledge

Bioterrorism and cosmetic surgery? A purified form of the botulinum toxin, called Botox, is actually used to help reduce lines and furrows in the skin. The solution is injected into the areas where the lines and wrinkles have formed. It works by preventing muscles that normally contract in a certain area underlying a line, crease, or furrow from acting. This makes the overlying skin temporarily smooth and wrinkle-free. The effect is not permanent and the injections must be given approximately every six months. Side effects can include muscle weakness and discomfort, bruising, drooping eyelids, and light sensitivity.

Treatment and Recovery

Therapy for botulism consists of medical and nursing care as needed, including possible use of a ventilator for breathing. There is also therapy with an antitoxin—a chemical that blocks the action of the botulinum toxin (poison) in the blood. The antitoxin is not given to infants. Early intervention can minimize subsequent nerve damage and severity of disease, but it cannot reverse paralysis that has already taken place. In the United States, botulinum antitoxin is available from the CDC via state and local health departments.

Paralysis usually improves after several weeks, although full recovery can take much longer. If the toxin causes respiratory (breathing) failure, the patient can die.

Infectious Knowledge

The Biological and Toxin Weapons Convention is an international agreement that obliges nations not to develop, produce, stockpile, or acquire biological agents or toxins that are not used for peaceful purposes; related weapons and means of delivery are also covered under this agreement. Currently, 144 countries have signed the treaty. However, the effectiveness of this treaty was questioned in 2001, when the United States rejected a draft protocol designed to strengthen the inspection and enforcement provisions of the treaty.

Keeping the Terror at Bay

The best way to thwart bioterrorism is to do research on the major organisms, make sure our public health system is prepared to handle an attack, and encourage cooperation among nations to eliminate offensive weapons programs.

The Least You Need to Know

- Bioterrorism has been used throughout history to help win battles and wars.
- Many countries have programs in bioweapons research and development.
- The anthrax attacks in the United States in the fall of 2001 showed that bioterrorism is indeed a real threat.
- Other organisms that cause diseases, like tularemia and botulism, can potentially be used as bioterrorist weapons.

Chapter 21

Drug-Insensitive Bugs: Antibiotic Resistance

In This Chapter

- ◆ The emergence of antibiotic resistance
- ◆ How bugs resist drugs
- ◆ Making the problem worse
- ◆ What we can do about it

The discovery of antibiotics in the 1930s and their introduction into widespread use was the most important intervention against infectious diseases ever. Until that time, infectious diseases had been the primary cause of death. Although infectious diseases are still the major cause of death worldwide, in the United States, the leading causes of death today are heart disease and cancer.

We Won the Battle, but What About the War?

The antibiotic era began in the 1940s with the introduction of penicillin into general use. Researchers soon developed many other drugs effective in treating bacterial infections, and it seemed that the result would be victory in the war with infectious diseases. Public health officials became so confident of the power of antibiotics that they began to focus their research efforts on other areas.

Unfortunately, our confidence was premature. A powerful resurgence of infectious diseases began in the late 1970s. A significant increase in patients who are especially susceptible to infection has helped speed the spread of certain diseases and increased their life-threatening consequences. These groups include …

- Immune-compromised patients with HIV and some cancers.
- Patients on anti-cancer chemotherapy.
- Premature infants.
- People who have recently undergone organ transplants, heart surgery, and other invasive procedures.

Antibiotic Resistance

Perhaps the single most alarming component in the resurgence of infectious diseases is their increasing resistance to many, and in some cases all, known antibiotics. Disease caused by multi-drug-resistant bacteria poses a direct health hazard to all of us. Immune-compromised patients, patients going into the hospital for surgery, hospitalized patients requiring intensive care, the elderly in nursing homes, and children in day care centers are at especially high risk because they are more likely to be exposed to resistant organisms. These bacteria also threaten the prolonged survival of people with chronic lung disease, diabetes, and kidney disease. In most of these patients, sooner or later bacterial infection occurs and it is often the ultimate cause of death.

Many drug-resistant infections are treatable, but in some cases, especially with hospitalized patients, by the time doctors find drugs that work, the infection is too far advanced to save the individual.

> **CAUTION**
>
> **Antigen Alert**
>
> New drugs are not being developed quickly enough to replace those that are losing their effectiveness, and there are only a few programs of aggressive drug development in the works.

The scientific and medical communities are well aware that antibiotic resistance presents a major public health and scientific challenge. Almost a decade ago, scientists were calling for stronger partnerships between treating physicians and public health officials to improve our ability to identify and fight resistant infections. The danger is that if we do not preserve the effectiveness of existing drugs and develop new drugs, we may end up in a post-antibiotic era in which incurable infectious diseases run rampant.

The Ins and Outs of Antibiotic Resistance

Bacterial resistance is not a new phenomenon. By 1944, penicillin resistance in staph bacteria, the most common cause of hospital infections, was well-established (see Chapter 22 for more on hospital infections). By 1960, most staph was resistant to penicillin. By 1980, staph were becoming resistant to the next generation drug—methicillin. More recently, drug-resistant tuberculosis was a problem in New York City in the early 1990s, when one strain infected patients and health-care workers in a large number of city hospitals.

The reality is that bacteria are constantly evolving, and they have demonstrated an ability to develop sophisticated ways to resist most antibiotics we've developed so far.

A bacterium's response that leads to drug resistance is part of a survival mechanism, finely tuned over millions of years, that enables bacteria to persist and thrive despite constant chemical and biological challenges in hostile environments like the soil. All the wonders of modern drug discovery pale next to the diversity of mechanisms available to ensure survival.

Biochemical Modes of Resistance

Bacteria have developed several ways to resist the drugs we use to treat them. Once the drug is in a patient's body, bacteria sometimes are able to destroy the drug, or to modify it just enough so it cannot do its job. Some bacteria can block access to the target—the place the antibiotic needs to get to. Others can alter the antibiotic target site so that when the drug gets there, it's ineffective. There are even bacteria that have developed biological "pumps" that push the antibiotic out of their cells, rendering it useless!

Biological, professional, and societal factors have all contributed to the emergence of drug resistance.

Biological Factors

Biological resistance can emerge in one of two ways.

- The first is that as bacteria multiply, there are a certain number of random mutations, or changes, in bacterial DNA. Every once in a while, one, or a combination, of these naturally occurring mutations will make bacteria drug-resistant.

- The second way is that some organisms exchange DNA with other organisms. The

Antigen Alert

The excessive use of antibiotics in the dairy and poultry industries has helped to flood the environment with antibiotics. This helps to select and encourage the growth of environmental organisms that are strong enough to resist the presence of low-level doses of antibiotics.

new DNA is integrated into the chromosome and is copied when the organisms multiply. If that exchanged DNA contains genes that confer resistance, the organism that got the DNA and its progeny will be drug-resistant, too.

Professional Factors

It is estimated that close to 50 percent of all antibiotic prescriptions written for patients are unnecessary. This over-prescription and overuse of antibiotics helps to kill weaker organisms in our bodies and encourage the survival and growth of stronger ones, which are capable of resisting the drugs.

Paradoxically, medical advances have also made us more vulnerable. Invasive procedures like open-heart surgery and organ transplants temporarily weaken our immune systems, making us more vulnerable to infection. Anti-cancer chemotherapy also makes us temporarily weaker and more vulnerable. Even though our immune system usually recovers, we may get sick with repeated infections. In time, after recurrent drug therapy, a drug-resistant infection may emerge and take hold.

Potent Fact

Should you use antibacterial soaps and cleansers? Many of the bacteria that live on our skin are not harmful and in many cases are even beneficial. The overuse of antibacterial soaps may not only kill some of these helpful bacteria, but they may also help contribute to further growth of resistant bacteria.

Infectious Knowledge

The hospital cost to treat a patient with a nondrug-resistant strain of tuberculosis is $12,000, versus up to $250,000 to treat a patient with a multi-drug-resistant strain.

Societal Factors

Patients who go to the doctor and insist on an antibiotic prescription—even when the doctor tells them their infection is not bacterial—also help to create an environment in their bodies that makes resistant infections more likely.

In addition, failure to control the spread of HIV, TB, and other epidemics adds to societal vulnerability.

Homelessness, intravenous drug use, poor hygiene, poor nutrition, poor sanitation, and overcrowding also provide a breeding ground for resistant bacteria.

Hospital Infections

We will cover specific hospital infections in the next chapter, but it's important to point out that a large number of hospitalized patients develop infections they did not have before they were admitted. Approximately one in three of these infections are antibiotic resistant. Patients who recover from these infections require an average of ten extra days in the hospital. The danger of drug-resistant hospital infections is exacerbated by the increasing number of patients, such as those undergoing long-term chemotherapy, those getting organ transplants, or those who are HIV-positive, whose immune systems don't function normally.

What Can We Do?

Collaborations linking the public and private sectors are necessary to combat the problem of antibiotic resistance. First and foremost, antibiotic-resistant infections must be closely monitored and tracked to identify drug susceptibility patterns, to develop DNA fingerprint profiles, and to clarify trends in the spread of resistant strains. New approaches are needed to disseminate information quickly, to encourage fast response. In order to accomplish this, there must be cooperation among laboratory scientists, clinicians, professional societies, public health officials, pharmaceutical companies, and others.

Be Vigilant with Antibiotics

Specifically, we need better ways to track antibiotic usage in hospitals and in the community. Part of this can be done through good public health practices, but we each have a personal responsibility to limit our use of antibiotics when they are not necessary. This starts with listening to our doctor if he or she tells us our symptoms are caused by a virus, not a bacteria. Antibiotics don't work against viruses, so if we insist on getting a pill anyway, we are adding to the problem.

Physicians need to take the time to try to educate their patients about the proper use of antibiotics. This goes beyond telling them when they are needed and when they aren't. Patients need to understand that if they are prescribed an antibiotic for 10 days, they need to take it for the full 10 days *even if they feel better after three days.* Why? Because after three days, a good number of the bacteria causing the infection are still alive in the body. In fact, antibiotics kill the weakest bugs first, so if you stop taking an antibiotic before you should, the stronger bacteria will survive; if you get sick again, you might end up with an infection that's drug-resistant. Depending on the type of infection, you might also be able to spread your resistant bacteria to others, making the problem that much worse.

> **CAUTION**
>
> **Antigen Alert**
>
> Once a bacterium becomes resistant to a drug or drugs, it and all of its progeny will remain resistant. There is no way to turn it back into a sensitive bug!

Speedy Diagnoses

The development of faster and more efficient diagnostic measures is also an important element in the fight against drug-resistant diseases. If an infection is quickly diagnosed and tested for drug susceptibility, then patients can be treated properly with drugs that will work. Because of the lag time between testing and results, many doctors begin treatment based on the symptoms they see and then modify the treatment when the tests come back. In most cases this practice isn't serious, but it does allow resistant infections to remain contagious longer, putting the patient and others at risk.

There are a number of tests in development that could determine both the cause of an infection and its drug susceptibility in a matter of hours, instead of the days or weeks that most tests now take. The use of these new techniques will be vital in our quest to keep potent some of the antibiotics that we use often today.

Bug Surveillance and Tracking

The development and use of new molecular techniques to track deadly strains of organisms is also important. The use of DNA fingerprinting to track tuberculosis strains is a good example.

In New York City in the early 1990s, there were a number of outbreaks of multi-drug-resistant tuberculosis. The Public Health Research Institute (PHRI) Tuberculosis Center DNA-fingerprinted tuberculosis strains from these patients, compared the fingerprints using a computerized matching program, and found that almost all the patients were infected with the same exact strain—the W strain—that was resistant to almost all antibiotics. Unfortunately, this work was done after many of the patients had already died. The value of this type of work for real-time surveillance and tracking is only beginning to be understood and appreciated. This particular type of fingerprinting can be done in a week, which is a lot faster than traditional forms of drug susceptibility testing for TB.

> **Infectious Knowledge**
>
> Over the past 10 years, the spread of the W strain of TB has been tracked around the world. In addition to finding it in the United States, France, and China, researchers have found that the strain is prominent in the epidemic of multi-drug-resistant TB infecting Russian prisons.

Develop New Drugs

A vital piece of an effective strategy will have to include new drugs. This requires basic research into the mechanisms of drug resistance so that we can modify existing drugs so they continue to work, as well as the development and testing of new antibiotics. The major obstacles to this kind of work include both a lack of funding and a perceived lack of return on investment on the part of the pharmaceutical companies. In other words, it costs hundreds of millions of dollars to develop new drugs, and if the market isn't there to sell them because it's not big enough or the people who are sick can't afford to buy the drug, the pharmaceutical companies can't make money. Furthermore, if the drug's use has to be restricted to prevent resistance from arising, sales can't be maximized.

Unfortunately, many drug companies are getting out of the antibiotic development business. It costs hundreds of millions of dollars and takes several years to develop new drugs. In the end, there just isn't much money to be made. Most companies are focusing on developing drugs for chronic conditions that patients need to take for long periods of time, if not a lifetime, for conditions like arthritis or high blood pressure, or quality-of-life drugs like Rogaine and Viagra.

Further understanding of how infection occurs and what makes certain bacteria, and certain strains of bacteria, in particular, more likely to cause infections is also an important piece of the puzzle.

Get Organized!

Finally, success in this fight will depend on our ability to organize hospitals, public health officials, government officials, citizens groups, and others to educate, legislate, and participate in a variety of activities that will help keep the antibiotic era alive!

The Least You Need to Know

- ◆ Antibiotic resistance began to develop soon after antibiotics were introduced into clinical practice in the 1940s.
- ◆ There are a variety of medical, social, and economic reasons that resistance has grown and flourished.
- ◆ Once bacteria are drug-resistant, they never go back to being drug-sensitive.
- ◆ We need better education, better medical practices, and a wide variety of public health measures to combat the resistance problem—along with new and better drugs.

Kicking Patients When They're Down: Hospital Infections

In This Chapter

- ◆ The nature of hospital-based infections
- ◆ How hospital infections spread
- ◆ The dangers of staph and strep
- ◆ Controlling and preventing hospital infections

Most people think of hospitals as places where sick people go to get better. Usually, this is true. Sometimes, though, people actually get sick *from* the hospital. Five to ten percent of all patients admitted to hospitals get an infection they didn't have before they checked in. These infections lead to longer hospital stays, higher costs for medicine, and sometimes even death! To make matters worse, more and more of these infections are resistant to the antibiotics we traditionally use to treat them.

More people die every year from hospital infections (90,000) than from accidental deaths (70,000), including motor vehicle crashes, fires, burns, and falls.

It's estimated that hospital infections cost our health-care system more than $4.5 billion dollars each year. And that is only part of the picture—hospital infections cost the economy many lost workdays and can turn a routine trip to the hospital into a nightmare.

The New Breed of Uber-Bacteria

Nosocomial infections can be caused by many different organisms that live in hospitals. The most common causes of hospital infections are bacteria, but other organisms, such as fungi, can cause infections as well. The organisms that cause hospital infections are sometimes different from the ones that cause similar infections in the community. In hospitals, organisms evolve using mechanisms that help them survive even in the presence of antiseptics and antibiotics. This makes them difficult to treat effectively. Even if the doctor recognizes the symptoms, it is impossible to tell which drugs will be most effective until test results that show which, if any, drugs will kill the bugs are available. This testing takes at least 24 to 48 hours.

> **Disease Diction**
>
> A **nosocomial** infection is the name given to an infection a person gets while they are in the hospital. It is derived from the Greek words *nosos*, which means disease, and *komeion*, which means to care for, as in an infirmary or hospital.

Windows of Opportunity

The super bacteria found in hospitals take advantage of special opportunities unwittingly created by advances in medical care and the sterile nature of the hospital environment itself. Some of the reasons these special infections occur include the following:

♦ The introduction of modern, invasive medical procedures like open-heart surgery and transplants have allowed doctors to keep sicker patients alive longer. The sicker a patient is, the more susceptible he or she is to infection due to a weakened immune system.

♦ Medical procedures themselves also make patients temporarily vulnerable. Prolonged chemotherapy temporarily suppresses the immune system, making cancer patients particularly vulnerable to infections. Within a few months after treatment, most cancer patients' immune systems return to normal. Transplant patients' immune systems are deliberately suppressed to prevent organ or tissue rejection.

> **Antigen Alert**
>
> More than 2 million patients a year are affected by hospital infections, at a cost of more than $4.5 billion per year. It's estimated that hospital infections are responsible for 50 percent of all complications that occur with hospitalized patients.

◆ The increased number of people who are HIV-positive and have weakened immune systems give infections more opportunities to spread.

The CDC estimates over 1.5 million cases of nosocomial infection in long-term-care facilities and nursing homes occur each year, or an average of one infection per year per patient.

◆ As a result of illnesses and bacteria that patients bring with them when they are admitted, hospitals are home to many different infectious organisms. Some of them can stay alive on surfaces of equipment such as catheters and gurneys, whereas others live on people—on their skin, hands, or even in their nostrils. Consequently, health-care workers and visitors can carry and spread infections.

Antigen Alert

Infections are monitored and reported in two ways:

◆ **Incidence** is the number of new cases within a given time period. It shows how quickly a disease is spreading.

◆ **Prevalence** is the absolute number of cases in a given population—either at a point in time or over a period of time. Unlike incidence, prevalence includes both old and new cases, so it shows the impact of a disease on a population.

How Hospital Infections Spread

Hospital infections are frequently transmitted through person-to-person contact. They can be spread from the hands or nose of a health-care worker to a patient, and then perhaps to another patient. They can also be spread to patients from infected equipment, which is why sterilizing equipment each time it is used is essential. Finally, if the organism is highly infectious, one patient can spread it to others in the same room or ward.

Antigen Alert

Methicillin-resistant *Staphylococcus aureus* (MRSA) is a drug-resistant strain of staph and a frequent cause of nosocomial infection, posing serious problems for infection control. Burn victims are particularly susceptible to MRSA infection because their skin, which normally serves as a barrier to infection, has been damaged. Extended hospitalization and antibiotic therapy have been identified as additional risk factors for MRSA infection.

A recent study over an 18-month period at a major New York hospital involving more than 100 patients indicated that MRSA is difficult to eradicate, even when its presence is known. Surveillance, epidemiological studies, and the introduction of strict infection-control regimens can reduce the prevalence of MRSA but may be insufficient for eradication or prevention of outbreak situations.

As noted in Chapter 1, when a particular organism spreads to a group of patients, it is called an outbreak. In a hospital, outbreaks can occur when a group of patients are infected by a health-care worker who is a carrier of a particular organism. Many people carry staph bacteria in their noses, but are not sick themselves.

When a Sick Patient Gets Even Sicker

If a patient shows symptoms of a hospital infection (specific infections and their diseases are covered later in this chapter), his or her physician will order a culture right away. Meanwhile, though, the doctor will probably start giving the patient antibiotics based on the symptoms. In 24 to 48 hours, the doctor will know which organism is causing the infection and what drugs will work to treat it. If there is a need to change the drugs the patient is already taking, the doctor will do so.

> **Infectious Knowledge**
>
> Most hospital infections are treated with oral drugs, but some more severe infections are treated intravenously, often because the patient has an intravenous line and/or cannot take medication by mouth.

Patients whose infections are highly contagious are put in isolation rooms to prevent the spread of the disease to the rest of the hospital population. Anything that comes in contact with these patients is sterilized before being used again. Doctors, nurses, and anyone else who sees the infectious patient will wear masks and gloves to prevent further spread of the infection.

Although many hospital infections are curable, some patients die from them. Recently, sportswriter Dick Schaap died of an infection he got in the hospital after surgery for a hip replacement.

Hospital Infection Enemy #1: *Staphylococcus*

The most common cause of hospital infections is the *Staphylococcus* bacteria, more commonly known as staph. Staph bacteria can live harmlessly on the skin, especially around the nose and mouth. If the skin is punctured or broken, though, the bacteria can get into the wound and cause infection.

Staph can attack any part of the body, from the skin, eyes, and nails to the inner lining of the heart. The symptoms depend on where the infection develops, but can include boils (large, pus-filled swellings most frequently seen on the face, neck, buttocks, or armpits), carbuncles (clusters of boils under the skin), swelling and redness around a cut or sore, swollen lymph nodes, pneumonia, kidney infection, and damage to the heart. Patients who are already sick, cancer patients, AIDS patients, and intravenous drug users are especially susceptible to severe infection and more likely to die if they get a drug-resistant strain of the disease.

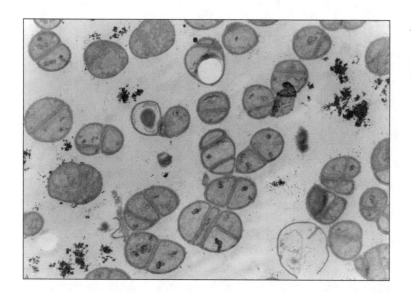

An electron micrograph of Staphylococcus, the bacteria that is the most common cause of hospital staph infections.

Staph infections sometimes spread through the bloodstream to the bones and joints. Left untreated, they can cause permanent stiffness and arthritis. If staph gets into the lungs, patients can get life-threatening pneumonia. And if staph attacks the heart, permanent damage can occur to that vital organ.

Potent Fact

Doctors and nurses should always change their gloves between patients.

Sticking It to Staph

Staph infection is diagnosed based on the appearance of symptoms and a staph culture, which takes between 24 and 48 hours. The culture indicates whether the infection is staph and whether or not it is a drug-resistant strain. As with most infections requiring lengthy diagnoses, many doctors will begin treatment based on symptoms before the test results come back.

The most common antibiotic used to treat staph infections is methicillin. However, many strains of staph are now resistant to this powerful drug. If a patient has a methicillin-resistant staph infection, there is only one drug left that will cure it—a drug called vancomycin. Physicians and researchers are concerned that the staph bugs will eventually become resistant to vancomycin as well—strains with reduced susceptibility to vancomycin have already been detected in Michigan, New Jersey, New York, and Illinois. If that happens, staph infections could run rampant in hospitals unless a more powerful antibiotic is developed.

Potent Fact _____

Bacteria can become resistant to drugs by acquiring resistance genes from other bacteria that are already resistant to the drugs. These genes "jump" from one species to another. In hospitals, where infectious organisms come in contact with one another frequently, there is the potential for all sorts of gene jumping, and thus the development of new drug-resistant strains. For example, the *Enterococcus* bacteria (discussed later in this chapter) is already resistant to vancomycin, and physicians fear that jumping genes from the *Enterococcus* bacteria to the staph bacteria will cause staph to become resistant to vancomycin as well. Researchers have shown this can happen in the laboratory, but fortunately, it hasn't happened in hospitals yet.

The Flesh-Eating Bacteria: Group A Strep

The Group A type of *Streptococcus* bacteria can cause a variety of diseases, including the rapid destruction of large areas of body tissue—hence the name flesh-eating bacteria. Group A strep also causes strep throat, scarlet fever, and less harmful skin infections.

The so-called flesh-eating disase is an infection of the skin and deeper tissues, which is characterized by extensive and rapidly spreading death of tissue and underlying structures. It starts out when the bacteria enter the body through a small or even invisible cut in the skin. The first symptom is mild redness. Within three days, there is swelling, the skin turns dusky, then purplish, and bloody blisters develop. Without immediate treatment, the infection will spread all over the body—fast!

Infectious Knowledge
Jim Henson, founder and creator of the Muppets, died from a Group A strep infection in 1990.

That immediate treatment involves surgery to remove all dead tissues and antibiotics. Twenty-five percent of people who contract this disease will die from it.

The disease was first identified in 1884; fortunately, it is quite rare.

Enterococcal Infections

Enterococci are bacteria that normally live in the gastrointestinal tract and can cause a variety of systemic infections. Many enterococcal infections are resistant to even the strongest antibiotic—vancomycin—making them very difficult to treat. Although vancomycin-resistant enterococcal infections cause high death rates, data suggests that enterococcal infections that are susceptible to vancomycin are just as deadly.

Symptoms of enterococcal infections include the following:

- General symptoms like fever, chills, and muscle pain
- Swelling, redness, pus, or abscess around a surgical wound
- Changes in mental status

Doctors treat enterococcal infections on a case-by-case basis. Sometimes they try to use older drugs, a combination of drugs, and even experimental drugs in order to kill the bacteria.

Antigen Alert

Residents of long-term-care facilities are especially vulnerable to enterococcal infections because the bacteria thrive and spread when there are groups of sick patients in one place.

The Power of Soap and Water

Controlling hospital infections is a major task, often requiring full-time infection control teams. These teams consist of microbiologists, infection control doctors, and infection control nurses and scientists who monitor infections and respond to specific cases or outbreaks as needed. Their job is to quickly identify the source of an infection and how that infection is spreading so that they can stop it.

Surprisingly, the most effective way to prevent the spread of hospital infections is a simple one: hand-washing. It's especially vital for health-care workers to wash their hands between attending to each and every patient.

Doctors and nurses should always wash their hands with soap and water between patients, be sure to maintain sterility of equipment, and wear gloves. Many hospitals keep a box of sterile gloves in every room, making it easy for physicians and nurses to change them between patients.

Infectious Knowledge

Frequent hand-washing is the best way to cut down the number of health-care workers spreading infection to other health care workers and to their patients.

Infectious Knowledge

Dr. Joseph Lister, a surgeon, was one of the first to use antiseptics to control hospital infections. He used antiseptic sprays on open surgical wounds, helping to protect against infection. It just so happened that the solutions used in the sprays made doctors' hands raw and sore, so they started wearing rubber gloves during surgery to protect them from further irritation. As a bonus, they later discovered that gloves also cut down the risk of infection for patients!

Tackling a Difficult Problem

Although hospital infections cost hospitals, insurers, and patients millions of dollars, not enough is being done to prevent and control them. Several obstacles lie in the path to better prevention and control of hospital infections, including the following:

◆ **Prohibitive costs in a competitive market** As noted earlier in this chapter, hospital infections are very costly to insurers, patients, and hospitals themselves. However, because the cost of controlling and preventing infections cannot be billed directly to patients or patients' insurance companies and because hospitals are constantly looking for ways to cut costs to stay competitive, infection-control budgets are often the first to be cut.

◆ **Lack of agreed-upon standards** Currently, there are no agreed-upon methods for tracking hospital infections. Each hospital might use its own method of documenting and tracking infections, making it difficult to compare the quality of care across hospitals and to get an accurate picture of the severity of the problem.

◆ **Difficulties in diagnosis** Hospital infections are not always easy to identify. Unfortunately, sometimes the first accurate diagnosis occurs at autopsy. Even then, clinical findings and lab test results must be analyzed by trained infection-control practitioners in order to identify and control potential problem organisms.

Potent Fact

Even the type of medical procedures a hospital does needs to be taken into account when documenting hospital infections, because some hospitals, such as teaching hospitals, do more cutting-edge, invasive procedures that leave patients temporarily vulnerable to infection. Transplant units of hospitals may also have higher infection rates than the rest of the hospital.

This means that although some hospitals may have higher infection rates, that may not mean they are bad hospitals, it may reflect more on the number and sophistication of the procedures they perform.

Identifying Infections

Tracing the source of hospital infections requires sophisticated techniques. Often there are many "flavors" of a particular species of bacteria found in a hospital at one time. Identifying the flavors, or strains, that are causing an outbreak and differentiating them from those that are harmless is key to controlling infection.

One method of typing is based on drug susceptibility patterns. This technique is being used along with more sophisticated DNA typing—molecular fingerprinting. DNA finger-prints look like supermarket bar codes; if fingerprints done from cultures from different patients look exactly the same, it's possible that an outbreak has occurred. If they look different, it's likely that the infections are cropping up individually and are not part of an outbreak. Further discrimination can be done through genetic sequencing, although the equipment to do this is still quite expensive, so fewer hospitals have access to this technique.

Another way to identify hospital infections faster is the use of molecular techniques for rapid diagnosis. There are methods available that take several hours to identify the cause of an infection, instead of a day or two, and some of these can determine drug susceptibility as well. Products to allow such testing to be done at the patient's bedside are under development.

Hospitals can also do cultures of all their employees to determine which ones are carriers of bacteria such as staph. Often, if someone is a carrier, the bacteria can be eradicated from their system with antibiotic therapy. Unfortunately, very few hospitals do this on a regular basis, and testing is not mandatory.

The Least You Need to Know

- Five to 10 percent of all patients who enter the hospital get an infection they did not have before admission.
- Many hospital infections are resistant to antibiotics.
- Hospital infections cost our economy billions of dollars per year in longer stays, expensive drugs, and lost work time for patients and their families.
- Although many patients are cured, some die of hospital infections.
- Having all health-care workers wash their hands and change their rubber gloves between patients will help reduce the incidence of hospital infection.

23

Diagnosing the Future

In This Chapter

- ◆ Improving our disease-fighting techniques
- ◆ Preparing for new epidemics
- ◆ Tracking, surveying, and diagnosing disease
- ◆ The promise of genome sequencing
- ◆ Remembering the basics

Despite the optimism of the antibiotic era, infectious diseases are still a serious threat to us all. As political and technological advances continue to put people from all over the world in close contact with one another, the need for coordinated, international efforts to combat infectious diseases—especially those that are spread through the air—increases. Furthermore, the threat of bioterrorism is much more real after the events of fall 2001, and we need to do whatever we can to be prepared.

Our public health system must have the resources—financial, medical, and scientific—to rapidly identify the cause of an outbreak or epidemic and to treat victims appropriately. There must be coordination on all levels: local, state, federal, and international. Furthermore, public and private organizations must support cutting-edge research so that new technologies, drugs, diagnostics, and vaccines are discovered, developed, and made available quickly.

Improving Public Health Measures

For most of us, good public health programs are basically invisible. Our water is clean, safe, and drinkable. Garbage does not pile up on the streets. We vaccinate our children on a schedule suggested by CDC and local health departments. Public health organizations track diseases such as rabies and tuberculosis without most people being aware of it.

> **CAUTION**
>
> **Antigen Alert** _____
>
> In the 1970s when money was tight and tuberculosis cases were on the decline, the public health infrastructure that had been developed to deal with the disease was dismantled. Because tuberculosis often has a long incubation period and is spread through the air, this ended up being a shortsighted decision.
>
> In the early 1990s there were a series of outbreaks of multi-drug-resistant tuberculosis in New York City hospitals and shelters. Many of those who got sick and died were HIV-positive, but while doctors and nurses were caring for them, more than 200 of them were exposed. All these people now carry a highly resistant strain of the bacteria and if they get sick, will have drug-resistant TB that may infect others. New York City spent more than $1 billion in response to the outbreaks of the early 1990s rebuilding an infrastructure to deal with a growing problem, including building hospital rooms with negative air pressure, refurbishing a TB hospital on Governor's Island where patients with drug-resistant TB could be incarcerated until cured, and hiring public health staff to implement directly observed therapy programs so that patients could be monitored to be sure they took their medicine. These programs were effective, but if they are dismantled again, it will only be a matter of time before we have another outbreak of TB. It is wiser to keep at least a skeleton program in place than to dismantle it and have to start from scratch the next time the inevitable outbreak occurs.

In the fall of 2001, when anthrax was delivered through the mail to United States senators and television news stations, public health officials and organizations suddenly became more visible. Thousands looked to the CDC for guidance and answers. They wondered, "What do you do if you think you've been exposed to anthrax?" "Should you get your doctor to write a prescription for Cipro?" "Why did some people get Cipro and others get doxycyline?" "Which drug is better and why?" "How rapidly can a potential anthrax exposure be confirmed?"

Despite conflicting opinions about how the anthrax cases were handled and how government agencies did or did not communicate with one another, there's no doubt that the public health system played an important role—without them, the outbreak might have been much worse. To be better prepared when the next attack or epidemic occurs, these organizations need better funding and support.

Communication Is Key

The United States has the most extensive disease surveillance and response network in the world. But there are gaps in its ability to detect outbreaks early.

In the case of West Nile virus in 1999 (see Chapter 14), it took a long time to establish the link between dead birds and sick people. If there had been established communication links between those who deal with human health and those who deal with animal health, West Nile probably would have been identified much earlier.

When a new or unidentified disease appears, our public health system must be able to identify the disease, work across state, local, and federal lines to determine how widespread it is, and quickly implement an effective control plan.

First and foremost, this requires good communication. Turf issues and jurisdictions must take a back seat to finding out what is going on and containing it. Public health officials need to be willing to work across disciplines with physicians, researchers, and veterinarians.

Policies, planning, and training are needed to ensure that early detection and surveillance are effective. Even the most effective drugs and vaccines will be ineffective if we don't identify which ones we need. When policies are developed at the federal level, they must be shared with state and local officials so everyone is aware of them in case of disease outbreaks or bioterrorist attacks.

One of the most important things we need to do is to improve laboratories' capabilities. For example, only a few labs in the U.S. have the equipment and staff capable of performing diagnostic tests rapidly enough to deal with new and emerging diseases or bioterrorist attacks. We must upgrade our laboratories, taking advantage of new technologies that are being developed by top academic researchers around the world.

> **Infectious Knowledge**
>
> In addition to overall disease surveillance, a networked, nationwide program to conduct surveillance on the prevalence of hospital-acquired infections, particularly those that are antibiotic-resistant, is extremely important. Our ability to preserve the effectiveness of existing drugs and determine how to modify them or make new drugs to treat resistant infections depends on effective surveillance.

Surveillance Is Key

According to the World Health Organization (WHO), surveillance of infectious diseases is crucial for prevention and control. *Surveillance* is defined as the ongoing systematic collection, collation, analysis, and interpretation of data; and the dissemination of information to those who need to know so that action may be taken.

Surveillance is used to ...

◆ Monitor disease trends.

◆ Monitor progress to see if methods used to control infections are working.

◆ Estimate the size of a health problem.

◆ Detect outbreaks of an infectious disease.

◆ Evaluate intervention and preventive programs.

◆ Identify research needs.

We can use the powerful computers now available to conduct disease surveillance programs and share data electronically. Ten years ago, scientists in New York and the Netherlands developed a database of "most-wanted" tuberculosis strains that were causing outbreaks in New York and elsewhere. A number of countries shared information about strains in their midst, and this gave those countries that were not experiencing outbreaks an opportunity to track their existing tuberculosis cases and identify a potential outbreak in its infancy. This began with the use of DNA fingerprinting technology to characterize different TB strains and a computer-matching program to compare the fingerprints.

The CDC developed a voluntary network that now includes more than 100 hospitals to share information on hospital infections. Unfortunately, with data from hospitals scattered throughout the country and reported infrequently, it is difficult to make judgments and develop policies regarding hospital infections. However, this type of program is a step in the right direction.

Certain diseases, such as tuberculosis, are reportable. In other words, any time a doctor or hospital finds a case, they are required to contact their local or state health department. Reportability can be a tool for conducting effective surveillance.

> **Disease Diction**
>
> **Surveillance** is the ongoing systematic collection, collation, analysis, and interpretation of data; and the dissemination of information to those who need to know so that action may be taken.

> **Infectious Knowledge**
>
> Each state has lists of diseases that are reportable both by laboratory directors and health-care providers. Common reportable diseases include chlamydia and other STDs; tuberculosis; Lyme disease; anthrax; viral hepatitis A, B, and C; HIV; botulism; and rabies.

Rapid Diagnosis Is Key

One of the most important elements of an effective public health program—one that can contain, control, and prevent disease—is the ability to rapidly identify the cause of an illness. Such identification involves determining the kind of organism causing the disease—is it a bacteria or a virus, for example—and whether it's susceptible or resistant to antibiotics. This information is crucial for figuring out how to treat sick patients, whether isolation is necessary, and who else around the patient should be tested for exposure to the disease.

As we have said in earlier chapters, most diagnostic tests still depend on looking for the presence of antibodies—the immune system's response to attack by an external invader—in the blood. It usually takes between 24 and 48 hours to discern antibodies. In many cases, such tests are appropriate because the diseases aren't all that dangerous and early identification is not a life and death matter.

At times, though, speed is of the essence. Rapid identification of a resistant hospital infection, within hours, for example, could go a long way to both treating the patient and preventing the infection from spreading. Rapid identification of a tuberculosis infection, which can take weeks to diagnose, could help to slow the spread of this airborne organism.

Public health officials fear that future bioterrorist attacks could spread a genetically altered organism the likes of which we've never seen before. It could be a hybrid combination of a bacterium and a virus, or a bacterium genetically altered to produce a potent toxin from another bacterium. By using molecular techniques to diagnose an illness caused by such an organism, public health officials will be better able to identify the disease-causing organism, and thus take appropriate measures to stop the spread of the disease and treat people already infected with it.

Infectious Knowledge

Using a PCR-based diagnostic method, researchers have developed a technique to confirm the presence of anthrax—either from a patient or from an inert surface, like an envelope, desk, or computer keyboard—within hours. It is in the early experimental stages, but could be available for health department and hospital use within a year. If it works for anthrax, the same technique will be used to develop diagnostics for other organisms that might be used in a bioterrorist attack.

To date, these new technologies have not been widely available to government and public health laboratories because they are in development and can be costly. It is important, however, that all the technologies and techniques that can be used to prevent the rapid spread of disease be made available to those who need them most, including public health and hospital laboratories. The days saved by not having to send a sample across the country for identification could save hundreds of lives.

Disease Diction

A **gene** is the smallest unit of heredity. A **genome** is all the genes of a living organism. **Genomics** is the analysis of genomes in order to understand how genes function and therefore how organisms work.

Infectious Knowledge

All diseases have a genetic component. Some diseases are inherited—caused by the genes we get from our parents. Some are caused by genetic mutations. In others the genetic component comes into play when the body responds to infection. For example, genetic differences can sometimes explain why some people are more vulnerable to certain infections while others are able to fight them off with ease.

The Impact of Genome Sequencing

One of the most important advances in molecular biology in recent years has been the sequencing of genomes of a number of organisms, as well as the sequencing of the human genome. Sequencing is the determination of the exact order of all the base pairs of DNA in an organism. Often specific genes are sequenced, too. How does this impact our approach to infectious diseases?

Medical researchers used to study an organism one gene at a time. With the availability of whole *genomes*, the entire genetic makeup of an organism, it's possible for researchers to study the whole organism as a system—looking at many or all of its genes at once. This method can help in the development of better preventive measures like vaccines, better diagnostics, and better treatments.

For example, the genomes of some infectious diseases, such as staph and tuberculosis, have already been sequenced. Researchers can determine how genes function in these organisms more quickly and, hopefully, find genes that are vital to the life of the organism—those are good targets for developing drugs that kill them.

Drug Development and the Human Genome

Researchers hope that the human genome will help them develop a new generation of drugs based on genes. Researchers will use gene sequence and function information to develop new drugs, instead of relying on the traditional, which basically involved trial and error. The drugs will be targeted to work at specific sites in the body, and should have fewer side effects than many of today's medicines.

Disease Diction

Pharmacogenomics is the study of how a person's individual genetic makeup affects his or her body's response to drugs.

The relatively new field of *pharmacogenomics* is the study of how a person's individual genetic makeup affects his or her body's response to drugs. It is possible that over time, drugs could be custom-made for individuals and adapted to each person's specific genetic makeup. Environment, diet, age, lifestyle, and state of health all can influence a person's response to medicines, but understanding a person's genetic makeup is thought to be the key to creating personalized drugs with greater safety and effectiveness.

Other potential benefits of pharmacogenomics include …

- **More powerful medicines**   Drug companies and researchers may be able to develop drugs based on the actual proteins, enzymes, and RNA molecules associated with genes and diseases. This will allow them to produce medicines targeted to specific diseases. It is hoped that this accuracy will increase the drugs' effectiveness while decreasing damage to nearby healthy cells.

- **Better, safer drugs the first time**   In time, doctors will be able to look at a patient's genetic profile and prescribe the best available therapy for that patient. This will help to lower the impact of side effects.

- **More accurate methods of determining appropriate dosages**   Dosages today are based on the patient's weight and age. In the future, this will be replaced with dosages based on their genetic makeup. This will include how well the body processes the medicine and the time it takes to metabolize it. This will maximize the drug's effectiveness and decrease the likelihood of overdose.

- **Advanced screening for disease**   Having your genetic makeup available will allow you to make appropriate lifestyle and environmental changes at an early age, to lessen the severity of a genetic disease. Knowing susceptibility in advance will also allow treatment to be used at the right time to maximize its effectiveness.

- **Better vaccines**   Vaccines made of genetic material, either DNA or RNA, will provide all the benefits of existing vaccines with lower risk. They will activate the immune system but they won't cause infections. They will be inexpensive, stable, and easy to store.

Researchers and public health officials also hope that pharmacogenomics will help speed the approval process for new drugs. Clinical trials can be developed to target patients who are most likely to respond to a particular drug, reducing the risk of a part of drug having catastrophic results.

If researchers can more quickly develop better drugs and determine what drugs will work most effectively on patients, it is possible that overall health-care costs will be reduced.

Gene Therapy

Gene therapy, or using people's own genes to treat disease, is still in the very early stages of development, but researchers are excited by its possibilities. It might be possible to use gene therapy to treat or even cure genetic and acquired diseases by using normal genes to replace or supplement a defective gene, or to bolster immunity to disease. For example, a gene could be modified to help stop tumor growth or to bolster immunity against certain disease-causing organisms.

Disease Diction

Gene therapy is a new approach to treat, cure, or prevent disease by changing the way a person's genes work.

Gene therapy is still in the experimental stages. Most experiments with it involve attempting to change a patient's genome to help them overcome a disease. This change would not be passed on to the next generation—the person's children—but there is early work being done on gene therapy that would target egg and sperm cells, with the goal of passing on the healthier genes to the next generation.

The Importance of Basic Research

With all the excitement surrounding genomics and the new rapid molecular diagnostic methods, it's important to understand and appreciate the vital role that basic medical research plays in the advancement and improvement of health care around the world.

Researchers at private organizations, government agencies, pharmaceutical, and biotechnology companies figure out how microorganisms cause disease; find targets for drug, vaccine, and diagnostic development; and determine how disease-causing organisms become resistant to drugs. They study our body's immune responses and determine which of our genes are activated and how they work when we get sick.

The federal government provides significant support for researchers through the National Institutes of Health (NIH) and other agencies, including the Department of Defense. It is vital that this support continues to be a national priority and that new programs, such as bioterrorist-preparedness programs, include significant research components to help us stay a few steps ahead of the bugs.

Potent Fact

President George W. Bush has proposed a $27.3 billion budget for the National Institutes of Health (NIH) for 2003. This is part of a five-year doubling of the NIH budget that began in 1998.

The government must also encourage public-private collaborations and cooperation among research and academic organizations so that new advances are shared, to improve the health of people worldwide.

The bugs that cause disease have been around for 3.5 billion years. People have been around for a tiny fraction of that time, so the odds are the bugs will continue to evolve and outsmart our efforts to destroy them. We must be vigilant and continue to learn more about the bugs in order to figure out how we can coexist with them so they don't destroy us.

The Least You Need to Know

♦ Improved public health systems and programs are vital to our ability to fight infectious diseases.

♦ Good disease surveillance helps to identify, contain, and control new and existing infectious diseases.

♦ Rapid diagnostics will be key to identifying new diseases and possible bioterrorist attacks. Technologies that can diagnose diseases in hours are currently under development.

♦ The science of genomics can play an important role in future drug, vaccine, and diagnostic development and has the potential to revolutionize modern medicine.

♦ Basic research still serves a vital function and needs our support.

Disease Diction: A Glossary

acellular vaccine   A vaccine made from the part of an infectious organism that produces antigens. The flu vaccine is an example of this.

Acquired Immune Deficiency Syndrome (AIDS)   A disease caused by infection with the human immunodeficiency virus (HIV). AIDS occurs when an infected patient has lost most of his or her CD4 T cells, an important component of the immune system's ability to fight infection, so that infections with opportunistic pathogens occur and do damage.

active TB infection   A person with active TB has symptoms including cough, chest pain, coughing up blood or sputum, weakness, fatigue, loss of appetite, chills, fever, and night sweats. They can spread TB to others. They have a positive skin test, and they may also have an abnormal chest x-ray and/or positive sputum smear or culture.

acute infections   Usually have a rapid onset and require medical attention to cure.

adaptive immunity   The portion of the immune response that develops when the body is exposed to various antigens and builds a specific defense for each antigen to which it is exposed. Also called acquired immunity.

adjuvant   A substance that is added during vaccine production to increase, or boost, the body's immune response.

aerosolization   When tiny particles of a disease-causing organism become airborne and cause disease in those who breathe them in. Hantavirus and tuberculosis are spread this way.

amino acids   Organic compounds that are essential components of the protein molecule.

anthrax   A soil organism that commonly infects domesticated and wild animals, such as cattle, sheep, horses, mules, and goats. People become infected when they come into contact with diseased animals, which includes their flesh, bones, hides, hair, and excrement. For this reason, anthrax has been a common source of disease in the textile industry, called wool sorters' disease. It's found sporadically throughout the world. Anthrax infection results in three different types of diseases: cutaneous (skin), gastrointestinal, and inhalation disease.

antibiotic   A substance that is produced by organisms such as fungi and bacteria, effective in the suppression or killing of microorganisms, and is widely used in the prevention and treatment of bacterial diseases. Antibiotics do not have any affect against diseases caused by viruses.

antibiotic era   Time period beginning in the 1940s with the introduction of penicillin into general use. This was followed by discoveries of many drugs effective at treating bacterial infections, and it seemed that the result would be the complete conquest of infectious diseases.

antibody   Any of various proteins in the blood that are generated in reaction to foreign proteins, or antigens, and are capable of neutralizing those proteins. Neutralization involves either killing the pathogens or preparing them for uptake and destruction by white blood cells. By neutralizing the proteins, the antibodies produce immunity against certain microorganisms or toxins. Antibodies are very specific.

antigen   Antigens are proteins, nucleic acids, carbohydrates, or other molecules from an infecting organism that the immune system recognizes and responds to with the creation of antibodies.

antitoxin   An antibody formed in response to and capable of neutralizing a biological poison.

attenuated vaccines   Vaccines made with a live organism that has been weakened by aging it or altering its growth conditions. Vaccines made this way are the most successful, probably because they grow in the body and cause a large immune response. They also have the highest risk of causing disease. Examples are measles, mumps, and rubella vaccines. Immunity usually lasts a lifetime, so no booster shots are required.

B cells (B lymphocytes)   One of the two major classes of lymphocytes that produce antibodies to fight infection.

bacteria   Plural of bacterium.

bacteriocidal   An agent that destroys bacteria.

bacteriostatic   An agent that prevents bacteria from growing, enabling the body's immune system to function more effectively.

bacterium   Any of numerous one-celled microorganisms occurring in many forms existing either as free-living organisms or as parasites, and having a broad range of biochemical, often disease-causing properties.

biological warfare   *See* bioterrorism.

bioterrorism   The intentional use of disease-causing organisms—bacteria, viruses, or fungi—as weapons to cause serious illness, resulting in disabling physical and psychological trauma and/or death.

bone marrow   The site of generation of the cellular elements of blood, including red cells, monocytes, leucocytes, and platelets. The bone marrow is the site of development of

B cells and stem cells that give rise to T cells, important components of the immune response to infection.

botulism   A bacterial infection characterized by paralysis of nerves on both sides of the body that starts in the head and moves downward. Blurred vision and impaired speech are common initial symptoms. Symptoms of food-borne botulism may begin as early as two hours or as long as eight days after ingestion of toxin. Typically, cases are present 12 to 72 hours after the infected meal.

bubonic plague   An infection caused by the bacterium *Yersinia pestis*, which occurs in wild rodents and is transmitted to people. Symptoms include high fever, chills, muscle pain, headache, swollen lymph glands (buboes), and seizures. It can be treated with antibiotics.

Camphylobacter   A bacterium that causes food-borne illness. It normally lives in the intestines of mammals and warm-blooded birds. It can survive refrigeration and grows if food is left out for too long at room temperature. The organism is sensitive to heat, so proper cooking and pasteurization will kill it.

carrier   A person who is colonized with a disease-causing organism has that organism living on their skin or in their body. Carriers often are not sick, but can spread disease if they come in contact with others, such as hospital patients.

chickenpox   A very contagious disease that is caused by a virus. More than 95 percent of Americans get chickenpox by the time they reach adulthood. Symptoms include an itchy rash that starts in the scalp and spreads to the stomach, back, and face. The disease is spread from coughs and runny noses, or direct contact with the fluid from sores of someone who is sick.

chlamydia   Both the name of the bacterial cause and the name of the most common sexually transmitted disease (STD) in the United States today. It can infect the penis, vagina, cervix, urethra, or eye. Symptoms include abnormal discharge (mucus or pus) from the vagina or penis or pain while urinating. Early symptoms may be very mild but usually appear within one to three weeks after infection. Often, people with chlamydia infections have few or no symptoms of infection and fail to get treated.

cholera   A disease that causes profuse watery diarrhea that leads to rapid dehydration and, if not treated, death. Serious disasters such as hurricanes, typhoons, or earthquakes increase the risk of contracting cholera among area residents. The cholera bacteria cannot survive in an acidic environment, so people taking antacids or other products to reduce stomach acid are more susceptible.

chronic infection   May start slowly and last for a long time. Often medical help manages a chronic infection but cannot cure it.

clone   A population of genetically identical cells produced by divisions of one original cell.

commensalism   "Eating at the same table." A neutral situation where the host and the organism live together, but have no effect on each other's life cycles. This is the relationship humans have with most microorganisms, particularly bacteria.

cowpox   The common name of the disease produced by vaccinia virus, used by Edward Jenner in the successful vaccination against smallpox, caused by the related variola virus.

Creutzfeld-Jakob disease (CJD)   A rare, fatal brain disease that causes rapid, progressive dementia and associated neuromuscular disturbances. It's a horrible disease that leads to a gruesome death. It's the human form of mad cow disease.

cutaneous   Occurs and stays on the skin.

cytokines   Proteins made by cells that affect the behavior of other cells. Cytokines act on specific cell receptors and thereby induce new activities in the cell, such as growth, differentiation, or death.

dengue   Also called dengue hemorrhagic fever, causes death in more than 20 percent of cases of the disease when medical attention isn't available. That percentage can be reduced to less than one percent, however, with modern medical intervention. The same mosquito that carries the yellow fever virus can carry the dengue virus as well.

dermatophytes   Name for a group of fungi that commonly cause skin disease.

diphtheria   A bacterial disease that invades the throat. It's spread through contact with salivary or nasal secretions, such as coughs and sneezes, from someone who is infected.

directly observed therapy (DOTS)   When a health-care worker meets with the patient every day, or several times a week, to be sure they take their medicine.

dysentery   An inflammation of the intestine characterized by the frequent passage of feces with blood and mucus. Like cholera, dysentery is spread by fecal contamination of food and water, usually in impoverished areas with poor sanitation.

E. coli   A bacterium that lives everywhere in the environment. In people and animals it lives in the digestive tract. The *E. coli* bacteria get into the stomach and small intestines and attach themselves to the inside surface of the large intestine. Most *E. coli* is beneficial, secreting vitamins K and B complex in our gut. But some strains of *E. coli* can cause diseases. In disease-causing strains, toxins or poisons the bacteria secrete cause swelling of the intestinal wall, which is what makes people sick.

earache   Common childhood problem often caused by upper respiratory infections. The ear aches because of inflammation of the middle ear, caused by fluid that builds up behind the eardrum. Kids often cry, pull on their ears, have a fever, act irritable, and are unable to hear well.

Ebola   A virus found in some African countries that causes the Ebola disease. In humans, it is transmitted through the blood and causes severe damage to the liver, lymphatic system, kidneys, ovaries, and testes. Platelets and linings of arteries are severely damaged, which results in profuse bleeding. Mucosal surfaces of the stomach, heart membrane, and vagina are also affected. Internal bleeding results in shock and acute respiratory distress, leading to death.

effector cells   Lymphocytes that can mediate the removal of pathogens from the body without the need for further differentiation, as distinct from naïve lymphocytes, which must proliferate and differentiate before they can mediate effector functions, and memory cells which must differentiate and often proliferate before they become effector cells.

ELISA   Enzyme linked immunoassay test. A blood sample is taken and tested for the presence of a specific antibody or antigen.

encephalitis   Swelling of the brain. It can be caused by viruses and bacteria, including those transmitted to people by mosquitoes.

endemic   Prevalent in or peculiar to a particular locality.

enteric infection   One that affects the intestines.

epidemic   When an infectious disease spreads beyond a local population, lasts longer than a simple outbreak, and reaches people in a wider geographic area.

epidemiology   The study of epidemics. Where they happen, when they happen, what factors allow them to happen, and what can be done to contain and prevent them.

false negative   An incorrect test result reported when your body hasn't responded to an infection with the production of antibodies, so the test indicates there isn't an infection when in fact there is one.

false positive   An incorrect test result indicating the presence of infection when there isn't one.

fungus   Any of numerous plants lacking chlorophyll, ranging in form from a single cell to a body mass of branched filaments that often produce specialized fruiting bodies and include yeasts, moulds, and mushrooms. Fungi can cause a variety of diseases.

gene   The smallest unit of heredity.

gene therapy   The correction of a genetic defect by the introduction of a normal gene into bone marrow or other cell types.

genital herpes   Herpes Simplex Virus (HSV), better known as genital herpes, is a contagious viral infection. Two types of virus cause genital herpes: one commonly causes sores on the lips resulting in fever blisters or cold sores and the other can also infect the mouth.

genital warts   Human papillomavirus (HPV) causes genital warts, as well as cervical and other genital cancers. Genital warts are growths or bumps that appear on the vulva, vagina, anus, cervix, penis, scrotum, or thigh. They may be raised or flat, single or multiple, small or large, or clustered to form a cauliflower-like shapes.

genome   All the genes needed for a living organism.

genomics   Understanding genetic material on a large scale.

giardia lamblia   A one-celled, microscopic parasite (protozoa) that lives in the intestine of people and animals and passes in the stool of an infected person or animal. It is protected by an outer shell that helps it survive outside the body and in the environment for long periods of time.

giardiasis   Illness called by *Giardia* protozoa. Symptoms are diarrhea, abdominal cramps, bloating, weight loss.

gonorrhea   A curable sexually transmitted disease. The most common symptoms of gonorrhea are a discharge from the vagina or penis and painful or difficult urination. It can infect the genital tract, the mouth, and the rectum.

Guillain-Barré Syndrome   A rare nerve disease that causes paralysis. Sometimes full paralysis occurs and lasts for months. Often patients must be hospitalized in intensive care units for long periods of time. Full recovery is common, but some people are left with severe, permanent nerve damage.

hantavirus   A hemorrhagic virus that causes hantavirus pulmonary syndrome.

hantavirus pulmonary syndrome A disease caused by the hantavirus that starts out with fatigue, fever, and muscle aches. About half of the people who get sick also have headaches, dizziness, chills, nausea, vomiting, diarrhea, and stomach pain. Four to 10 days after the first symptoms appear, coughing and shortness of breath begin. This is due to the lungs getting filled with fluid.

Helicobacter pylori A rod-shaped bacteria that is found in the protective mucous layer of the stomach. About one half of the strains of the bacteria produce poisons that help induce gastritis and peptic ulcers.

helminths Multi-cellular parasites (worms and flukes) that have complex reproductive cycles that involve intermediate hosts for larval stages and a separate host for the adult form.

hemolytic uremic syndrome (HUS) Most common cause of kidney failure in children. It develops when bacteria get into the circulatory system through an inflamed bowel and releases certain toxins into the blood. Most cases occur after gastrointestinal illness, most commonly diarrhea.

hemorrhagic colitis The sudden onset of stomach pain and severe cramps caused by *E. coli* infection. This is followed by diarrhea that is watery and bloody. Sometimes there is vomiting, but there is no fever. The incubation period is three to nine days. The illness lasts about a week and there are usually no long-term problems.

hemorrhagic fever A fever characterized by severe bleeding, often from the mouth, stomach, and digestive tract.

hepatitis A disease of the liver that causes inflammation and swelling, potentially resulting in permanent damage. There are many strains of this disease.

hepatitis A Strain of hepatitis transmitted by human consumption of fecal-contaminated drinking water or food. Hepatitis A is the most common vaccine-preventable disease in international travelers.

hepatitis B Strain of hepatitis that may develop into a chronic disease in up to 10 percent of newly infected people each year. If left untreated, the risk of developing cirrhosis (scarring) and liver cancer becomes higher.

hepatitis C Strain of hepatitis that causes both acute and chronic liver disease. Unlike other types of hepatitis, more than 80 percent of hepatitis C infections become chronic and lead to liver disease.

hepatitis E Strain of hepatitis that has symptoms much like hepatitis A. It is an acute, short-duration disease spread widely in many tropical underdeveloped countries usually through contaminated drinking water. Hepatitis E affects young adults, rather than children, and causes a high death rate, particularly in pregnant women.

hepatitis G Strain of hepatitis transmitted by blood-borne routes. Risk groups include intravenous drug users, hemodialysis patients, and transfusion recipients. Acute infection is diagnosed by antibody tests or by detection of RNA in a person previously RNA negative.

hepatocellular carcinoma (HCC) The most common malignant tumor found in males worldwide; it is closely associated with hepatitis.

host   The living creature in which a disease-causing organism lives, multiplies, and causes disease.

human immunodeficiency virus (HIV)   An agent responsible for causing acquired immune deficiency syndrome (AIDS). HIV is a retrovirus of the lentivirus family that selectively infects CD4 T cells, leading to their slow depletion and corresponding depletion of the body's ability to fight infection.

immune response   Response made by the host to defend itself against a pathogen.

immune system   Name used to describe the tissues, cells, and molecules involved in adaptive immunity, or sometimes the totality of the host's defense mechanisms.

immunity   The ability to resist infection.

immunization   The deliberate provocation of an adaptive immune response by introducing an antigen into the body.

immunoglobulins   A family of plasma proteins including all antibodies.

incidence   The number of new cases within a given time period. It shows how quickly a disease is spreading.

infection   Invasion of body tissue by disease-causing microorganisms and their subsequent growth, production of toxins (poisons), and injury to that tissue.

infectious   Capable of causing and spreading disease.

infectious disease   An illness caused by a microorganism, virus, or other disease-causing entity that can be spread and cause serious illness or death.

inflammation   A general term for the local accumulation of fluid, plasma, proteins, and white blood cells that is initiated by physical injury, infection, or a local immune response. Acute inflammation describes early and often transient episodes, while chronic inflammation occurs when an infection persists or during the formation of antibodies.

interferons   Cytokines, or cellular chemicals, that can induce cells to resist viral replication.

killed or inactivated vaccine   A vaccination method in which the organism is killed and then injected into the body. The typhoid vaccine and the Salk polio vaccine are examples.

Koch's postulates   Evidence that a particular organism causes a particular disease, as defined by Robert Koch. There are four conditions that must be met: The organism must be present in every case of the disease, the organism must grow in a culture after isolation from the body, the cultured organism must cause the same disease if introduced into an uninfected host, and it must be recoverable from that host. Although these don't hold true for all diseases, they are still useful in diagnosis today.

latent tuberculosis (TB) infection   A person with a latent TB infection has no symptoms and does not feel sick. They cannot spread TB to others, although if they have a skin test it will indicate that they have been exposed to the bacteria.

leishmaniasis   A parasitic disease caused by a protozoa that initially lives in the sand fly and is transmitted to people through sand fly bites. The organism develops and multiplies in the gut of the fly and is introduced into the bloodstream of people after a bite. The disease occurs in three forms. In one, there are lesions on the skin, in the second, there are ulcers, and in the third, the systemic variety, there can be fatal complications.

leukocyte   General term for a white blood cell.

Listeria monocytogenes   A food-borne bacteria that causes infection in humans. It can cause stillbirth, meningitis, septicemia, or localized infections. The organism is found in soil, forage, and water. It can multiply even in refrigerated food.

Lyme disease   A chronic infection with *Borrelia burgdorferi*, a bacterium that can evade the immune response. It is transmitted to people by ticks. Its symptoms can be quite severe. They include skin problems, arthritis, muscle aches, fever, neurologic complications, and heart abnormalities.

lymph nodes   Organs where adaptive immune responses are initiated. They are found in many locations in the body where lymphatic vessels come together.

lymph system   A major component of the immune system. It is a complex circulatory system composed of a network of organs, lymph nodes, lymph ducts, and lymph vessels that produce and transport lymph (clear fluid containing white blood cells and antibodies) from tissues to the bloodstream.

lymphocytes   Class of white blood cells produced in the lymphoid organs throughout the body. These cells are responsible for cellular and humoral immune system actions. There are two classes of lymphocytes—B cells and T cells.

macrophages   Cells important in innate immunity, in early nonadaptive phases of host defense, as antigen-presenting cells, and as effector cells in humoral and cell-mediated immunity.

mad cow disease   A disease caused by rogue prions that fills cows' brains with holes and results in certain death. Although only cows can contract mad cow disease, humans can be infected with the human form, called Creutzfeld-Jakob disease, from eating tainted beef.

malaria   An infectious tropical disease caused by a protozoan transmitted to people by the bite of the female *Anopheles* mosquito. It is characterized by high fever, shaking, chills, sweating, and anemia. It may follow a chronic or relapsing course after initial illness.

Marburg disease   A hemorrhagic fever caused by the Marburg virus with symptoms very similar to ebola, although the fatality rate is lower, at approximately 25 percent.

measles   A disease caused by a virus. Early symptoms of measles include fever, runny nose, cough, and sore and red eyes, followed by the appearance of a red-brown blotchy rash on the skin. The rash usually starts on the face and spreads downward. It lasts at least three days. Although children who get measles can become fairly sick, most of them recover with no long-term negative effects.

microorganism   Microscopic organisms, unicellular except from some fungi, which include bacteria, yeasts, and other fungi and protozoa, all of which can cause human disease.

molecular genetics   The study of molecules (e.g., DNA and RNA) important in biological inheritance.

monocytes   White blood cells with a bean-shaped nucleus, which are precursors to macrophages.

mumps   Disease caused by a virus. Symptoms include low-grade fever and swelling or tenderness of one or more of the salivary glands in the cheeks and under the jaw. Pain is often worse when swallowing, talking, or chewing. Loss of appetite is also common.

nail fungus   Nail infections caused by fungus, which can manifest themselves in a variety of patterns. Sometimes a portion of the nail becomes thick and brittle. Other times, the fungi attack the cuticle and the growth spreads out from there.

natural killer cells (NK)   Cells that kill certain tumor cells. They are important in innate immunity to viruses and other intracellular pathogens.

nematodes   Parasitic filial worms that cause insect-borne infections.

nosocomial infections   Infections that are transmitted in hospitals. Some are opportunistic and cause disease because hospital patients are sick and weak. Others, like staph, may occur because of the nature of the hospital environment.

oncogenes   Genes involved in regulating cell growth. When they are defective in structure or expression, they can cause cells to grow continuously to form a tumor.

opportunistic infections   One that occurs when an organism causes disease by taking advantage of the opportunity provided by a person's weakened immune system. Often these organisms live peacefully in the environment or inside our bodies, but can turn deadly if conditions change.

outbreak   When a group of people in a small geographic area become ill with an infectious disease.

pandemic   An epidemic that has spread around the world.

parasites   Organisms that obtain sustenance from a live host. In medical practice, the term is restricted to worms and protozoa, the subject of parasitology.

pathogen   An agent that causes disease, especially a microorganism.

pathogenesis   The development of a diseased condition.

pathogenic   Capable of causing disease.

pertussis   Also whooping cough, a very contagious and dangerous respiratory infection caused by the *Pertussis* bacteria. The disease gets its name from the whooping sound children make when they try to breathe after a coughing spell. Symptoms of whooping cough generally include runny nose and a cough that gets worse and worse. Violent coughing spells can end with vomiting. Once the whooping stage begins, antibiotics don't work.

Peyer's patches   Aggregates of lymphocytes along the small intestine.

phagocytes   Macrophages that take up and destroy bacteria.

plasma   The fluid component of blood containing water, electrolytes, and proteins.

pneumocystis pneumonia (PCP)   The most common opportunistic infection in AIDS patients, developing in 60 to 85 percent of AIDS cases.

pneumonia   An inflammation or infection of the lungs. Air sacs fill with pus, mucous, and other liquid and can't function normally. It can be caused by many different organisms.

polio   A crippling viral disease with initial symptoms that include fever, fatigue, headaches, diarrhea, vomiting, constipation, stiffness in the neck, and pain in the limbs.

polymerase chain reaction (PCR)   A technique for amplifying a specific sequence in DNA by repeated cycles of synthesis driven by pairs of reciprocally oriented primers. It can be used for rapid identification of the cause of an infection and to determine whether that cause is drug sensitive or drug resistant.

postulate   Law.

prevalence   The absolute number of cases in a given population—either at a point in time or over a period of time. Unlike incidence, prevalence includes both old and new cases, so it shows the impact of a disease on a population.

prions   Proteins that exist in the brains of all mammals. There is evidence that some prions cause disease.

prophylaxis   The prevention of disease or a process which can lead to disease.

protective immunity   The resistance to specific infection that follows infection or vaccination.

protein   Any of a group of organic compounds that have amino acids as their basic structural units and that are found in all living matter and are required for the growth and repair of animal tissue.

protozoa   Single-celled animals, that may be aquatic or parasitic. The word protozoan comes from the Greek and means first animal.

rabies   Acute, usually fatal viral disease transmitted to people through the saliva of animals (usually dogs, cats, skunks, or raccoons).

Reiter's syndrome   A rare disorder characterized by arthritis and painful joint swelling. It may be venereal or it may occur after small bowel infections. It usually goes away within a year.

reportable   Designation for any disease that by law a doctor or hospital must report cases of to the local or state health department. Reportability can be a tool for conducting effective surveillance.

retrovirus   A virus that contains RNA as its genetic material.

river blindness   Caused by a parasitic worm that lives for up to 14 years inside the human body. Each adult female worm produces millions of larvae that migrate throughout the body and cause serious visual impairment and sometimes blindness, rashes, lesions, intense itching, depigmentation of the skin, elephantiasis, and general debilitation.

Rocky Mountain spotted fever   Caused by a bacteria that is spread to people through tick bites. Symptoms of the disease include fever, headache, and muscle pain, followed by the appearance of a rash. The disease is hard to diagnose in its early stages and can be fatal.

rubella   Also called German measles, a very contagious disease caused by a virus. The virus causes fever, swollen lymph nodes behind the ears, and a rash that starts on the face and then spreads to the torso and arms and legs. Rubella is not usually serious in children, but can be very serious if a pregnant woman becomes infected.

Salmonella   A common bacteria that causes a variety of intestinal infections. Symptoms usually show up in six to 48 hours. They include diarrhea, fever, and stomach cramps. Sometimes they start with nausea and vomiting.

scalp itch   A fungal infection of the scalp and hair that usually occurs in young children, but may appear in all age groups. It is contagious and may be spread from child to child in a school or day care setting.

schistosomiasis   Infection caused by five different species of waterborne flatworms called schistosomes. The worms enter to body through contact with infested water. Contact can be by washing hands, washing food, swimming, fishing, farming, and growing rice.

sepsis   Infection of the bloodstream. It is often fatal.

sexually transmitted disease (STD)   Any of a number of infectious diseases that are acquired through some type of sexual contact. Once acquired, it can be passed on to other sexual partners.

Shigella   A bacteria that thrives in the intestines and causes sudden, severe diarrhea. *Shigella* is spread in food, but it can also be spread from person to person. The symptoms of a *Shigella* infection are fever, stomach cramps, and diarrhea that is bloody and has mucous in it.

sickle cell anemia   A hereditary disease that affects molecules in the blood, resulting in sickle-shaped red blood cells.

smallpox   An infectious disease caused by the variola virus that once killed at least 10 percent of infected people. It has now been eradicated by vaccination.

sporozoite   A slender, spindle-shaped organism that is the infective stage of the malaria parasite. It is the result of the sexual reproductive cycle of the parasite, which occurs inside the mosquito.

staph   A bacteria that can attack any part of the body, from the skin, eyes, and nails, to the inner lining of the heart. These infections sometimes spread through the bloodstream to the bones and joints. Left untreated, they can cause permanent stiffness and arthritis. If staph gets into the lungs, patients can get life-threatening pneumonia. And if staph attacks the heart, permanent damage can occur.

subunit vaccines   Vaccines that use only the parts of an organism that stimulate a strong immune response. Researchers separate the disease-causing genes and then isolate and purify them to be used as a vaccine. The hepatitis B vaccine is an example of this.

surveillance   The ongoing systematic collection, collation, analysis, and interpretation of data about diseases; and the dissemination of information to those who need to know so that action may be taken.

syphilis   A sexually transmitted disease that begins with genital sores, progresses to a general rash, and then to disfiguring abscesses and scabs all over the body. In its late stages, untreated syphilis can cause heart abnormalities, mental disorders, blindness, other neurological problems, and death.

systemic infection   Any infection that gets into the bloodstream and can affect internal systems and organs of the body. Systemic infections are generally more serious than infections that remain on body surfaces like the skin or infections that are localized to a small area of the body.

T-cells   Subset of lymphocytes defined by their development in the thymus. They are a major component of cellular immunity. Some can kill invaders, others work with and influence the action of other white blood cells.

tetanus   Also called lockjaw, it is caused by a bacteria that is common in the soil. The bacteria die quickly when they are exposed to oxygen so any cut that isn't open to the air provides an environment where the bacteria can grow. For those not vaccinated, they can get tetanus by stepping on a dirty nail or getting cut by a dirty tool. The bacteria produce a toxin or poison that spreads in the bloodstream. The disease can result in severe muscle spasms, paralysis, and death.

toxin   A poisonous substance, having a protein structure, that is secreted by certain organisms and is capable of causing a pathological condition when introduced into the body tissues.

toxoid vaccines   Made from the poisons disease-causing organisms secrete. The toxins are chemically treated to decrease their harmful effects. Diphtheria and tetanus vaccines are toxoids. Since the immune response they induce can be weak, they are often given with an adjuvant—another agent that increases immune response.

tuberculosis (TB)   A disease caused by the bacteria *Mycobacterium tuberculosis*. The bacteria are passed from person to person through the air when someone with TB coughs or sneezes. People who are nearby may get infected after breathing in bacteria. The bacteria can attack any part of the body, but they usually stick to the lungs.

tularemia   Also called rabbit fever, a disease that can have several forms, but often includes fever, chills, and head and body aches. Complications include meningitis and severe pneumonia, among others.

ulcer of the stomach   Usually caused by a bacteria, not stress, and can result in burning or pain in the upper abdomen, and usually occur about an hour after meals or at night. The symptoms are often relieved temporarily by antacids, milk, or medications that reduce stomach acidity.

vaccination   The deliberate induction of adaptive immunity to a pathogen by injecting a dead or attenuated form of the pathogen.

variolation   Early process of inoculating people against smallpox by taking matter from the smallpox pustules of mild cases and scratching a person's arm or vein with them. The goal was to cause a mild infection of smallpox and stimulate an immune response that would give the person immunity from the natural infection.

vector   A free-living organism that carries an infectious organism and passes it to an animal host.

virulent   Extremely poisonous or extremely capable of causing disease.

virus   Any of various sub-microscopic disease-causing organisms, composed of a core of a single nucleic acid enclosed by a protein coat, able to replicate only within a living cell.

West Nile virus   The cause of West Nile encephalitis, a viral infection that is transmitted from mosquitoes to people.

Western Blot test   Diagnostic test. A Western Blot looks for specific antibodies that indicate infection with a specific organism.

yellow fever   Caused by a virus that is spread by a mosquitoes. Infection causes disease, with symptoms ranging from a mild flu to severe illness, including high fever, severe headache, muscle pain, jaundice, and vomiting of a liquefied black putrid matter. It is the pronounced yellow skin and eye color of severe jaundice that gives rise to its name.

Appendix B

References

Books

Armstrong, Donald A., and Jonathan Cohen. *Infectious Diseases*. London: Harcourt Publishers, 1999.

Farmer, Paul, et. al. *The Global Impact of Drug Resistant Tuberculosis*. Cambridge, MA: Harvard Medical School Program on Infectious Disease & Social Change and Open Society Institute, 1999.

Garrett, Laurie. *The Coming Plague*. New York: Penguin Books, 1995.

Janeway, Charles, and Paul Travers. *The Immune System in Health & Disease*. New York: Current Biology Ltd./Garland Publishing, 1997.

Kolata, Gina. *Flu*. Carmichael, CA: Touchstone Books, 2001.

Mandell, Douglas. *Principles & Practice of Infectious Diseases*. Gerald L. Mandell (Editor), John E. Bennett (Editor), Raphael Dolin (Editor). Churchill Livingstone, 4th edition, 1995.

Marquez, Gabriel Garcia. *Love in the Time of Cholera*. New York: Penguin Books, 1985.

Miller, Judith, et al. *Germs: Biological Weapons and America's Secret War*. New York: Simon & Schuster, 2001.

Preston, Richard. *The Hot Zone*. New York: Anchor, 1995.

Rosenberg, Charles E. *Cholera Years: The United States in 1832, 1849, & 1866*. Chicago: University of Chicago Press, 1987.

Shilts, Randy. *And the Band Played On: Politics, People, and the AIDS Epidemic*. New York: St. Martin's Press, 1988.

Tucker, Jonathan. *Scourge: The Once and Future Threat of Smallpox*. New York: Atlantic Monthly Press, 2001.

Turkington, Carol. *Hepatitis C: The Silent Epidemic*. Lincolnwood, IL: Contemporary Books, 1999.

Worman, Howard J., MD. *The Liver Disorders Sourcebook*. New York: McGraw-Hill, 1999.

Websites

General Health and Disease Information

www.cdc.gov
Centers for Disease Control and Prevention

www.cdc.gov/travel
CDC Traveler's Information

www.nfid.org
National Foundation for Infectious Diseases

www.niaid.nih.gov
National Institute of Allergy & Infectious Diseases

www.nih.gov
National Institutes of Health

www.nih.gov/medline
NIH Medline search

www.phri.org
Public Health Research Institute

www.who.org
World Health Organization

Disease-Specific Sites

www.aarc.org/resources/biological/history.asp
American Association for Respiratory Care

www.about–ecoli.com
Foodborne diseases

www.aegis.com/law/journals/1995/IOM95071.html
Law Journals—analysis of blood supply

www.ama–assn.org/special/hiv/support/support.htm
American Medical Association—HIV

www.ama–assn.org/special/std/treatmnt/guide/stdg3463.ht
American Medical Association—STD Treatments

www.caps.ucsf.edu/AIDSlist.htm
Center for AIDS Prevention Study, University of California San Francisco

www.cdc.gov/hiv
CDC Center for AIDS Prevention

www.cdc.gov/nchstp/od/nchstp.html
CDC Center for HIV, STD, and TB Prevention

www.gmhc.org
Gay Men's Health Crisis

http://history1900s.about.com/cs/malaria/
History of malaria

www.med.virginia.edu/hs–library/historical/yelfev/tabcon.html
University of Virginia

www.niaid.nih.gov/daids/vaccine/default.htm
NIAID Division of AIDS Vaccine

www.polioeradication.org/virus.html
Polio

www.thebody.com/amfar/amfar.html
American Foundation for AIDS Research

www.thebody.com/hotlines/national.html
National AIDS Hotlines and Resources

Articles

Abath, F.G., S.M. Montenegro, and Y.M. Gomes. "Vaccines Against Human Parasitic Diseases: An Overview." *Acta Tropica*, November 30, 1998, 71(3):237–54.

Alsahli, M., R.J. Farrell, and P. Michetti. "Vaccines: An Ongoing Promise?" *Dig Dis*, 2001, 19(2):148–57.

Arnon, S.S., et al. "Working Group on Civilian Biodefense. Botulinum Toxin as a Biological Weapon: Medical and Public Health Management." *JAMA*, February 28, 2001, 285(8):1059–70.

Balayan, M.S. "Epidemiology of Hepatitis E Virus Infection." *J Viral Hepat*, May 1997, 4(3):155–6.

Brinkman, K. "Evidence for Mitochondrial Toxicity: Lactic Acidosis as Proof of Concept." *Journal of HIV Therapy*, March 2001, 6(1):13–6. Review.

Crespo, A., and B. Suh. "Helicobacter Pylori Infection: Epidemiology, Pathophysiology, and Therapy," *Arch Pharm Res*, December 2001, 24(6):485–98.

Cuthbert, J.A. "Hepatitis A: Old and New," *Clin Microbiol Rev*, January 2001, 14(1):38–58. Review.

De Clercq, E. "Antiviral Drugs: Current State of the Art." *J Clin Virol*, August 2001, 22(1):73–89.

DeCross, A.J., and B.J. Marshall. "The Role of Helicobacter Pylori in Acid–peptic Disease." *Am J Med Sci*, December 1993, 306(6):381–92.

Dennis, D.T., et al. "Tularemia as a Biological Weapon: Medical and Public Health Management." *JAMA*, June 6, 2001, 285(21):2763–73.

Dowdle, W.R., et al. "Poliomyelitis Eradication." *Virus Res*, August 1999, 62(2):185–92. Review.

Enderlin, G., et al. "Streptomycin and Alternative Agents for the Treatment of Tularemia: Review of the Literature." *Clin Infect Dis*, 1994, 19:42–47.

Exhibit documents. Yellow Fever Epidemic of 1793. May 22, 1995, William L. Clements Library, The University of Michigan.

Franz, D.R., et al. "Clinical Recognition and Management of Patients Exposed to Biological Warfare Agents." *JAMA*, 1997, 278:399–411.

Gao, F., Bailes, et al. "Origin of HIV-1 in the Chimpanzee Pan Troglodytes." *Nature*, 1999, 397: 436–441.

Hayee, B., and A. Harris. "Helicobacter Pylori and Reflux Disease." *Lancet*, November 17, 2001, 358(9294):1730–1.

Health Aspects of Chemical and Biological Weapons. Geneva, Switzerland: World Health Organization; 1970:105–107.

Ho, D.D., et al. "Rapid Turnover of Plasma Virions and CD4 Lymphocytes in HIV-1 Infection." *Nature*, January 12, 1995, 373(6510):123–6.

Hoofnagle, J.H., and D. Lau. "New Therapies for Chronic Hepatitis B." *J Viral Hepat*, 1997, 4 Suppl 1:41–50. Review.

Hoofnagle, J.H. "Therapy for Acute Hepatitis C." *New England Journal of Medicine*, November 15, 2001, 345(20):1495–7.

Hull, H.F., and R.B. Aylward. "Progress Towards Global Polio Eradication." *Vaccine*, August 14, 2001, 19(31):4378–84. Review.

Isaacson, M. "Viral Hemorrhagic Fever Hazards for Travelers in Africa." *Clin Infect Dis*, November 15, 2001, 33(10):1707–12.

Knapp, J.S. "Historical Perspectives and Identification of Neisseria and Related Species." *Clin Microbiol Rev*, October 1988, 1(4):415–31.

Kottmann, L.M. "Pelvic Inflammatory Disease: Clinical Overview." *J Obstet Gynecol Neonatal Nurs*, October 1995, 24(8):759–67.

Lei, H.Y., et al. "Immunopathogenesis of Dengue Virus Infection." *J Biomed Sci*, September 2001, 8(5):377–88.

Leung, D.T., and S.L. Sacks. "Current Recommendations for the Treatment of Genital Herpes." *Drugs*, December 2000, 60(6):1329–52.

Liang, T.J., et al. "Pathogenesis, Natural History, Treatment, and Prevention of Hepatitis C." *Annals of Internal Medicine*, February 15, 2000, 132(4):296–305. Review.

Lind, I. "Antimicrobial Resistance in Neisseria Gonorrhoeae," *Clin Infect Dis*, January 1997, 24 Suppl 1:S93–7.

Markowitz, M., et al. "A Preliminary Evaluation of Nelfinavir Mesylate, an Inhibitor of Human Immunodeficiency Virus (HIV)–1 Protease, to Treat HIV Infection." *Journal of Infectious Diseases*, June 1998, 177(6):1533–40.

Marshall, B.J. "Helicobacter Pylori. The Etiologic Agent for Peptic Ulcer." The 1995 Albert Lasker Medical Research Award, *JAMA*, October 4, 1995, 274(13):1064–6.

Megraud, F., and B.J. Marshall. "How to Treat *Helicobacter Pylori*. First-line, Second-line, and Future Therapies." *Gastroenterol Clin North Am*, December 2000, 29(4):759–73.

Miller, L.H., et al. "Malaria Pathogenesis." *Science*, June 24, 1994, 264(5167):1878–83.

Morbidity and Mortality Weekly Report, October 19, 2001 / 50(41);893–7. Recognition of Illness Associated with the Intentional Release of a Biologic Agent.

Morbidity and Mortality Weekly Report Summary: CDC Anthrax Investigation. Updates and New Information. November 2, 2001/ Vol. 50/ No. 43 1–4.

Patz, J.A., et al. "Global Climate Change and Emerging Infectious Diseases." *JAMA*, January 17, 1996, 275(3):217–23.

Perelson, A.S., et al. "Decay Characteristics of HIV-1-Infected Compartments During Combination Therapy." *Nature*, May 8, 1997 May, 387(6629):123–4.

Purcell, R.H. "Hepatitis Viruses: Changing Patterns of Human Disease." *Proc Natl Acad Sci U S A*, March 29, 1994, 91(7):2401–6. Review.

Reed, W., et al. "Classics in Infectious Diseases. The Etiology of Yellow Fever: A Preliminary Note," The Philadelphia Medical Journal 1900, *Rev Infect Dis*, November/December 1983, 5(6):1103–11.

Reiter, P. "Climate Change and Mosquito–borne Disease." *Environmental Health Perspectives*, March 2001, 109 Suppl 1:141–6.

Ren, S., and E.J. Lien. "Development of HIV Protease Inhibitors: A Survey." *Prog Drug Res*, 2001; Spec No:1–34.

Richman, D.D. "HIV Chemotherapy." *Nature*, April 19, 2001, 410(6831):995–1001.

Robertson, S.E., et al. "Yellow Fever: A Decade of Reemergence." *JAMA*, October 9, 1996, 276(14):1157–62.

Shapiro, R.L., et al. "Botulism in the United States: A Clinical and Epidemiologic Review." *Annals of Internal Medicine*, August 1, 1998, 129(3):221–8.

Sharp, M., et al. "Origins and Diversity of Human Immunodeficiency Viruses." *AIDS*, 1994: S27–S42.

Stephenson, J. "Gene Mutation Link with HIV Resistance." *JAMA*, September 26, 2001, 286(12):1441–2.

Sutter, R.W., et al. "Poliomyelitis Eradication: Progress, Challenges for the End Game, and Preparation for the Post-eradication Era." *Infect Dis Clin* North *Am*, March 2001, 15(1):41–64. Review.

UNAIDS/WHO 2001 Joint United Nations Programme on HIV/AIDS (UNAIDS) World Health Organization (WHO) AIDS epidemic update, December 2001, pp. 1–23.

Vastag, B. "HIV Vaccine Efforts Inch Forward." *JAMA*, October 17, 2001, 286(15):1826–8.

Walker, U.A. "Clinical Manifestations of Mitochondrial Toxicity." *Journal of HIV Therapy*, March 2001, 6(1):17–21.

Wilson, M.E. "Travel-related Vaccines." *Infect Dis Clin North Am*, March 2001, 15(1):231–51.

Winstanley, P.A. "Chemotherapy for Falciparum Malaria: The Armoury, the Problems and the Prospects." *Parasitol Today*, April 2000, 16(4):146–53.

Wyler, D.J. "Malaria: Overview and Update." *Clin Infect Dis*, April 1993, 16(4):449–56.

Zullo, A., et al. "Helicobacter Pylori Infection and the Development of Gastric Cancer." *New England Journal of Medicine*, January 3, 2002, 346(1):65–7.

Index